T0267964

Is Anyone Listening?

Is Anyone Listening?

What Animals Are Saying to Each Other and to Us

Denise L. Herzing

The University of Chicago Press

Chicago and London

The University of Chicago Press, Chicago 60637
The University of Chicago Press, Ltd., London
© 2024 by Denise L. Herzing
Published 2024
Printed in the United States of America

33 32 31 30 29 28 27 26 25 24   1 2 3 4 5

ISBN-13: 978-0-226-35749-2   (cloth)
ISBN-13: 978-0-226-35752-2   (e-book)
DOI: https://doi.org/10.7208/chicago/9780226357522.001.0001

Library of Congress Cataloging-in-Publication Data

Names: Herzing, Denise L., author.
Title: Is anyone listening? : what animals are saying to each other and
    to us / Denise L. Herzing.
Other titles: What animals are saying to each other and to us
Description: Chicago ; London : The University of Chicago Press, 2024. |
    Includes bibliographical references and index.
Identifiers: LCCN 2024005296 | ISBN 9780226357492 (cloth) |
    ISBN 9780226357522 (ebook)
Subjects: LCSH: Human-animal communication. | Animal communica-
    tion. | Animal communication—Research. | Dolphins—Behavior.
Classification: LCC QL776 .H47 2024 | DDC 591.59—dc23/eng/20240301
LC record available at https://lccn.loc.gov/2024005296

♾ This paper meets the requirements of ANSI/NISO Z39.48-1992
(Permanence of Paper).

Contents

Preface

If you could ask a dolphin one question, what would it be? What might a dolphin ask you? These questions occurred to me when I first met a spotted dolphin in the wild one humid summer morning in 1985. I swam slowly away from my anchored boat through the gin-clear waters of a shallow sandbank in the Bahamas. The water was calm and peaceful, and there was no land in sight. Two dolphins approached and swam around me, looking directly into my eyes. Exchanging eye contact with a wild creature is like a splash of ice-cold water on your face. I sensed a keen and mutually exploratory awareness. Ten years later, after experiencing strong currents and large sharks, I would have a different type of respect for the ocean, one that wouldn't allow me to swim out so far alone with such a calmness. But this first experience was different.

There is nothing comparable to making contact with another intelligence and sensing that there is another "being" behind those eyes. In that moment, I wished I had studied to be an anthropologist; in all my years of working with whales and seals, nothing had prepared me for this. What is it like to meet and experience a new culture for the first time? What do you do if its members are curious and want to observe you?

My curiosity as a scientist about the lives of wild dolphins had taken me to the Bahamas, but it did not prepare me for this experience. As an animal of this world, I should have been better prepared: Our ancestors evolved with plants, animals, and our planet itself. We breathe the same oxygen, walk in the same woods, and feel the same wind and water as other creatures. I looked through a window to the dolphins' world, no longer estranged—as the land and sea seem in the open ocean—but intertwined, like a shoreline. We were mutually curious species, carefully considering each other.

Five years after that first encounter, I am on the Wild Dolphin Project's research boat with filmmakers and Sir David Attenborough. He's come to shoot a segment for a television series about the dolphins and their communication. We chat about some of the signals that dolphins use, and then a small group of adult spotted dolphins approach our boat. I've just told Sir David that sometimes I mimic the dolphins' little head nods to see if I can successfully turn the group in another direction. The group of dolphins swim by, and I offer to get in the water to bring the group back to the boat so he can meet them.

What was I even thinking to suggest such a thing? But I am committed to trying, so I enter the water, join the group of about eight adult male dolphins, and calmly swim with them for a while. Once I meet their eyes and appear to have their attention, I do my head nod to the side, signaling, as they do to each other, that I would like to turn to the right. Although I know I have no status in the dolphin's group, and no right to purposely change their direction of travel, I apparently do have their attention. The dolphin group turns with me, and as we approach the boat, I see Sir David getting ready to enter the water, and I notice the astonished faces of his film crew. Thinking like a scientist, I now doubt my agenda. Had I really learned to communicate with another species? Had they really turned because of me? I knew I couldn't determine the answers from a single interaction, but I felt lucky and maybe just a bit grateful. Even if I couldn't know why the dolphins had decided to visit the boat with me, the questions inspired by the interaction were at the core of my research. I agree with what Sir David said at dinner that night: "There is nothing that replaces the time and first-hand experience with an animal society."

Of course, I am not the first to wonder about interspecies communication and to understand the benefit of working with other animals as colleagues instead of subjects. We have learned about many social species, including chimpanzees from Jane Goodall, mountain gorillas from Dian Fossey, and African elephants from Cynthia Moss.[1] These pioneering researchers gave scientists like me examples for illuminating the lives of wild animal societies, while both observing and interacting. Early in my career, I decided to use their approaches for studying free-ranging dolphins.

Even in this, I am not alone; other researchers have explored animals' natural abilities to use signals. For example, research scientist Whitney Musser and her colleagues described how captive orcas (*Orcinus orca*), who usually prey on bottlenose dolphins, learned to use some of the dolphins' signals to socialize.[2] Robert Seyfarth, Dorothy Cheney, and Peter Marler clearly showed that vervet monkeys produce alarm calls that are seemingly specific to different predators.[3] And, similarly, in his compelling book *Chasing Doctor Dolittle*, Constantine "Con" Slobodchikoff described the detailed information provided in the alarm calls of prairie dogs, and even suggested we should identify those calls as language.[4]

In addition to describing my research, this book explains the work of other colleagues, and I hope it will offer a window into the questions that drive today's scientific exploration of animal language. I also hope to describe the process by which we study animal communication, from past to present. There have been many technological advances that now allow us to explore like we have never been able to before. As we look more closely into the minds of animals, we need to enter relationships as both speakers and listeners, learning what we can about other species.

In a sense, when it comes to interspecies communications, humans have a lot to learn from other animals. Research has shown that other species have evolved the cognitive abilities to listen, interpret, and decipher their neighboring species.

Charles Munn and other scientists have described how birds

(class Aves) of different species learn the alarm calls of their neighbors and use this information for their survival.[5] We know from biologist John Ford that although orcas normally communicate only with their own pod using its specific dialect, they share a mutual set of calls when interacting with another pod.[6] And sometimes it seems interspecies signaling functions both to communicate and to hide. This makes sense—we can imagine stealth communication with our allies that is indecipherable to our enemies. Ecologist Lydia Mäthger and colleagues have researched squids, octopuses, and other cephalopods who can simultaneously communicate secret messages to each other using polarized light invisible to their predators.[7] These cephalopods use chromatophores—skin cells that change colors—to communicate messages literally written on their bodies. They're not the only life-forms to use this bodily steganography; researchers including Hamish Ireland have even read between the lines to find communication messages in zebras' stripes.[8] These natural forms of encryption are a result of millions of years of evolution.

This book is also an opportunity to talk about the focus of my current work: using artificial intelligence (AI) tools—like machine-learning software that can help us categorize sounds—to look for language in animal communication. With today's discussion of bots or AI agents like ChatGPT, the concept of technology mimicking language will no doubt seem familiar. As a scientist who has worked for over three decades to implement similar systems with wild dolphins, I have collaborated with other people, across research disciplines, to create new communication programs with our unique dataset of dolphin sounds and to determine potential language patterns. In the last few years, I have worked with computer scientist Thad Starner and his team at the Georgia Institute of Technology, and we have been able to identify and analyze dolphin sounds in new ways.[9] I argue that we see rules, including grammar, that appear to be very important to the dolphins. We are creating a user interface so that other researchers can use these new tools to explore their recordings of other animals.

The most elusive part of this work, in the end, will be to interpret meaning. Does a mother use the same sequences of sounds to call her calf? Does this sequence only vary by the name she is calling? Our detailed dataset is allowing us to take a close look at this variation to interpret further dolphin signals. It is eye opening to be able to identify, as a computer can, the voice of a single dolphin, and yet I'm driven by the need to understand how entire groups of dolphins are conversing. Finally, with the development of our two-way computer system, CHAT (Cetacean Hearing Augmentation Telemetry), we may have a way to play back specific sequences of sound, in real time, to watch how the dolphins respond to their own language-like structures. And, as I discuss later in this book, you'll see that I question how to do this carefully and respectfully with potential dolphin interlocuters. In the end, this will be the ultimate test for whether you understand a language: can you listen, and can you talk back?

Among humans, there are many ways to communicate with those who don't speak the same language. In much the same way anthropologists and linguists have described human cultures and languages around the world, Hal Whitehead and Luke Rendell have shown, through their research, a variety of whale cultures—demonstrating that cultures exist in the myriad of nonhuman species on our planet.[10] Now, with these new tools and data, we are poised to see if language exists in other species. Whether whales and dolphins also have many ways to communicate with each other, across their cultures, is yet to be determined.

These ideas may seem familiar, as you may have encountered them in the findings of scientists—or you might have found them in the imaginations of science-fiction writers. Fans of the 1990s television show *seaQuest DSV* will remember the dolphin Darwin who, equipped with a backpack containing a special translation computer, advised the submarine's captain, played by Roy Scheider, and his crew. In the 2016 film *Arrival*, the challenge Amy Adams's character faces, as a human linguist, is to communicate with aliens to avoid a war. Can she understand these aliens' intentions before it's too late?

That film portrays some real methods that researchers use to explore the communication and cognition of other species. Fraught with difficulty and potential for miscommunication, deciphering a nonhuman being is challenging. An interesting twist in *Arrival*, which we will also explore in this book, is the idea that our reality is shaped by how we think and communicate. The Sapir-Whorf hypothesis, named after linguist Benjamin Whorf, suggests that the structure of a language determines or greatly influences the modes of thought and behavioral characteristics of the culture in which it is spoken.[11] In the movie *Arrival*, the nonprimate aliens have a sense of time that moves forward and backward, and so does their written language. We may find surprises as we explore nonhuman language that will similarly stretch our imaginations and skills.

What will encourage us to take up this challenge? What is the practical value for humans of communicating with animals? I often hear this question when I give public talks. I respond that it is important that humans understand the natural world, to better care for and save it. But, in a sense, even that seems self-serving. Instead, we might look to the ideas from philosophers, like Arne Naess, who have put forth the notion of deep ecology, that nature has a right to exist for itself.[12] I hope this book will encourage you to appreciate animal communication for its own sake, regardless of its applications for humans.

Back to those questions I asked in the Bahamas nearly forty years ago: *If you could ask a dolphin one question, what would it be? What might a dolphin ask you?* Whatever dolphins and other animals are telling us, I hope this book will encourage you to listen to it all—and to respond with inclusivity.

1

When Species Meet

Mimicry between species is perhaps the most fascinating of all.

Let's start with my first visit to the Bahamas. I was there for six weeks to determine whether I might observe a group of friendly Atlantic spotted dolphins in the wild. Equipped with an underwater video camera with a hydrophone, I would enter the water and try to not disturb the dolphins' behavior, keeping to myself while I recorded them.

The dolphins were cautious at first, eyeing me but keeping their distance. After the first few days, it was clear to me that this was an incredible opportunity to study this group in clear water and on a regular basis, something not easily accomplished in the open ocean. I went back to San Francisco with my life's plan. Given that most small dolphin species live on average twenty-five years, I estimated that twenty years would be my minimum time commitment to follow them through their lives and document their communication system. I was already focused on trying to crack the code of their communication signals and understand how this aquatic society dealt with its ocean world. Regardless of the outcome, I decided that I would spend a significant portion of my life actively listening to another animal.

Even though cracking the code of language is elusive, I have hope when I see humans working with other animals for practical or scientific reasons. Let's start with one of the classic examples of working together—finding food. Animal behaviorists Karen Pryor and Jon Lindbergh have shown how three generations of people who fish and bottlenose dolphins in Brazil work together—with the dolphins finding fish and the humans catching them.[1] Typically, dolphins help herd the fish and give cues, such as tail slaps and other body movements, to show the people where to put nets in the water. Together these two species—human and dolphin—easily round up the plentiful fish and share the bounty. Biologist Paulo Simões-Lopes and colleagues have shown similar dolphin-fisherman relationships in the Laguna region of Brazil,[2] and the relationship isn't exclusive to dolphins and people. Take for example the sub-Saharan hunters who follow honeyguide birds—also known as indicator birds—to bee colonies. Biologists Hussein Isack and Heinz-Ulrich Reyer have described how, when the birds guide humans, each species benefits: the birds provide clues to the hives' locations, and the humans knock down the hives, helping the birds access the honey.[3]

As we'll see throughout this book, scientists—hungry for a better understanding of animal behavior—have long worked with animal societies for a mutual goal of improving our knowledge. In the 1960s primatologist Jane Goodall pioneered the idea of considering an animal's "*Umwelt*"—or perception of the world—in her work with wild chimpanzees. She observed the chimpanzees at close range, recorded behavior to hypothesize trends, recognized individuals to follow their long-term changes and life histories, and studied chimpanzees in the context of their own community, while limiting habitat and behavioral disruption. Her work helped illuminate the chimpanzee as a highly intelligent and intensely social being capable of close and enduring attachments, and rich communication through sounds, gestures, and postures. In her writings, Goodall—a cultural observer in the chimpanzee world—describes the richness of chimpanzee life in the wild, including the politics, families, and struggles. Goodall recognized that the chimps she studied were individuals with dignity and different personalities. Although some scientists felt that animals should have numerical labels, Goodall gave them names. She wondered if, after she had been absent, the chimpanzees would forget her. But when she returned to them, they were even more tolerant of her presence. She also learned to use physical and vocal mimicry to establish a connection with her primate subjects over the years.

Like Goodall, the late primatologist Dian Fossey used mimicry in her study of the highland mountain gorilla. She described two different kinds of contact with the gorillas: one was pure observation, when the gorillas were unaware that she was watching them, and the second was when the gorillas were aware that they were being observed. Fossey encouraged the gorillas' curiosity in order for them to become more comfortable, suggesting that once their curiosity was satisfied, they resumed their usual activities. From the Kabara group of gorillas, she learned to accept the animals on their own terms and never to push them beyond the levels of tolerance that they were willing to give. In Fossey's early days, she would mimic the gorilla's chest beating as a way to get attention. But as she learned how important this act was for alarm and aggression, she decided to use it only when meeting new groups to get their initial attention. This gave her a strong conviction of research etiquette. "Any observer is an intruder in the domain of a wild animal and must remember that the rights of that animal supersede human interests."[4] As an observer, Fossey noted that a researcher must keep in mind that an animal's memories of one day's contact might well be reflected in the following day's behavior. Throughout her work, Fossey remained concerned about the highland mountain gorilla, an endangered species, in relation to human interaction, poaching, and habitat protection.

Conservationist Cynthia Moss has spent decades studying another long-lived social mammal, wild elephants. Elephants are very special animals: intelligent, complicated, and intense. Moss has watched them in their natural social groupings, which revolve around giving birth, mating for the first time, leaving the security of their families, and defending their young. Like other researchers, Moss takes pride in knowing individuals, having a relationship of trust, and understanding them as members of their society. Moss has found that, like humans, some elephants are curious and tolerant of strangers, while others are not.

With these lessons in mind, I was aware of the challenges of getting close and comfortable with another species. But, given dolphins live in the water, I wondered how I would get close enough to observe them while also allowing the dolphins to be comfortable enough to go about their daily lives. You can't really build an underwater blind, as terrestrial researchers might do. And you can't always mount a camera on the seabed since dolphins don't stay in the same place. I chose to habituate the dolphins to my presence in the water and tried to learn dolphin etiquette and limits, so I would not disturb their day-to-day lives.

I focused on simply getting underwater video of their behavior with correlated vocalizations. I achieved this by mounting an underwater microphone, a hydrophone, on my video housing so it would patch directly into the audio of the video camera. Even in my early years of research, I had already envisioned a computer that allowed me to input dolphin sounds and identify patterns. And I dreamed of software that would let me overlap video and sound for a fuller understanding of the dolphins' visual and sound communication. Animal behaviorists had been studying and recording animal behavior for decades, and the basics don't really change, just the tools. So, I began the extensive process of gathering long-term information about this dolphin species and, specifically, about this resident dolphin community. As the years and decades passed, I added what technology I could. But it was only after 2010, when powerful computer tools—specifically what's called machine learning, algorithms that can adapt and improve by following explicit human-given instructions for analyzing data—became readily available and modifiable for my rapidly growing video and sound dataset, that I really grew excited.

For food, for understanding, for data. There are many reasons we meet with other species. Combined with some of our many emerging computer tools, to help us with things that remain difficult for the human brain to analyze, human observations and insights also add richness to our eventual interpretation of what animals do when they communication, either with each other or with humans.

We have considered how humans can enter other animals' cultures, but animals are forced to adapt to our culture when humans decide to breed or capture them. As we've seen with Goodall's and Fossey's work, mimicry, or the imitation physical postures or acoustics, is in some ways a technique of reaching out to another species to enhance social bonds. Animals, too, reach out in captive settings through mimicry and attention-seeking behaviors, perhaps their way of trying to communicate in their strange new human world.

Although physical mimicry has been observed with primates and other mammals, vocal mimicry is rarer. But marine mammals, including pinnipeds (seals, sea lions, and walrus) and cetaceans (whales and dolphins), are known for their vocal abilities and, in the case of dolphins, vocal learning. Because of their aquatic environment, these mammals show extraordinary sound production and reception capabilities, and acoustic mimicry is often a natural trait found in their own societies.

Various researchers have described the spontaneous mimicry of human words by captive marine mammals. One example is Hoover, a harbor seal who lived in an outdoor pool at Boston's

Figure 1.1. A dolphin and a human mimic each other by laying on the sand. Dolphins sometimes mimicked our postures as well as our vocalizations in the water. Photograph courtesy of the Wild Dolphin Project.

New England Aquarium and who is described in detail by biologist Katherine Ralls and colleagues.[5] Visitors would pass by Hoover and other harbor seals in the pool on their way into the aquarium. At night unhoused people would reportedly visit or spend the night around the pool and talk to Hoover. Over time, Hoover seemed to learn to position his body vertically in the water and tuck in his very large chin, apparently giving him the vocal manipulation necessary to mimic human words.

I heard Hoover when I visited the aquarium during a marine mammal conference in Boston in 1989. "Hello there. Hello there." These were the first words I heard. "Hoover. Hoover. My name is Hoover." At first, I thought I was crazy, and then I was just fascinated. For some reason, this unique individual named Hoover had observed and interacted with humans enough to get their attention by mimicking these words—quite adeptly I must say. I remember watching as unsuspecting colleagues approached the pool and became wide-eyed with surprise as they heard a harbor seal talk.

Some toothed whales, including dolphins, belugas, and pilot whales, have also spontaneously mimicked humans or human-generated sounds. The late Sam Ridgway, known as the "father of marine mammal medicine" for his work developing dolphin anesthesia, and colleagues described a beluga whale who began to make humanlike sounds in a tank where he regularly heard humans talking to each other during tank cleaning and other activities.[6] Companionship and the need to connect, especially in captivity, can be a powerful driver for an intelligent animal. So, is it a surprise that one species may reach out to another to connect?

Dolphins who spontaneously mimic human and mechanical sounds have also been observed in the wild. Biologist Ana Alves and colleagues described a group of pilot whales who spontaneously mimicked naval sonar signals.[7] Probably more an insight into how these animals greet each other in a natural setting, the spontaneous mimic of a mechanical sound is an attempt at initiating contact, nevertheless. In my work with my team, over many decades with the same individual Atlantic spotted dolphins, there were multiple occasions where the dolphins mimicked our body postures (laying on the sandy bottom), our activities (swimming using a dolphin kick), or our human words as we greeted them through our snorkels (this took the form of a mimic of the duration of a word and an equal duration of one of their burst-pulse sounds).

Spontaneous mimicry of human sounds is not limited to marine mammals. Zoologist Angela Stoeger and colleagues describe an Asian elephant imitating human speech—actually altering how it produced sounds to more closely match human vocalizations.[8] And primatologist Serge Wich and colleagues describe how an orangutan spontaneously learned to mimic a human whistle and then learned to change features of that whistle, like duration and frequency, as needed during the interaction.[9]

Perhaps it should not be such a surprise that animals can mimic human sounds or activities to get social attention. When groups of animals synchronize and coordinate complex flight or swimming, these are only an external manifestation of all these other mechanisms within our bodies. Reaching out socially to other species may be a common ability among many species. Animals are not aliens—we have evolved in one world.

Marine mammals also encounter humans when they are stranded on beaches and need help from various rescue groups. In 2009, I had the opportunity to help with the release of a young male juvenile Atlantic spotted dolphin in Key West. Cutter, as he was named from his initial sighting around a Coast Guard cutter in port, was observed by marine mammal rescue teams and finally pulled out of the water by the Marine Mammal Conservancy, a rescue group based in Key Largo, Florida. Clearly out of place in the Key West harbor, and visibly emaciated and lethargic, Cutter was eventually rescued and taken to a rehabilitation facility in Key Largo, where he began his three-month rehabilitation. I was called to observe him and try to verify his age. Most researchers are used to working with bottlenose dolphins, both because of their coastal habitats and the fact that they are the species most found in captivity. So, when I estimated Cutter to be a juvenile Atlantic spotted dolphin, probably three or four, and quite capable of foraging on his own for the most part, members of the rehabilitation staff were surprised. They quickly came to understand that spotted dolphins are smaller than their bottlenose dolphin cousins, and what looks like a young bottlenose dolphin calf could be a juvenile spotted dolphin. Cutter had some spots on his underside, which, from my long-term observations, meant he was at least two to three years old and more likely three to four years of age. This was critical information since releasing a calf, due to their dependence on their mother for milk and protection, is not allowed by the National Marine Fisheries Service. We were given permission to try to rehabilitate Cutter for potential release. We knew there were many steps in this process: Cutter had to pass a medical check, show normal behavior, and be able to eat live prey. As well, we would only be given permission to release him if we found a group of spotted dolphins in the nearby waters for his reintroduction. Most rescue groups are careful about human contact during the rehab process, so animals don't get too comfortable around humans, but it's a double-edged sword. Dolphins are social and need attention and companionship, so it makes sense that humans should provide this to help keep them calm and heal. In my view, you aren't going to take away twenty-five million years of evolution in three months of human contact.

Luckily a local boat captain, Alma Armendariz, informed us that there had been a large group of Atlantic spotted dolphins seen on a regular basis since February, when Cutter stranded. My thoughts were that Cutter simply got lost, cut off from his group for some reason—maybe while fleeing for survival from a predator—or perhaps Cutter was just an adventurous juvenile. We thought if we could find this local group, Cutter might have a chance to go back into the wild. So, after all his check points, we had a release plan. The Wild Dolphin Project's research vessel, *Stenella* (a sixty-two-foot power catamaran), would be the reconnaissance boat down in Key West, looking for spotted dolphins in the area. Meanwhile, Cutter was transported to a small pool, awaiting the day when a boat would run him out to meet up with other spotted dolphins. Cutter's release was dependent on us finding others of his species, and since spotted dolphins are not commonly found near Key West, we hoped it would be his family. It was early May, and my team and I were due to start our field season in the Bahamas. But we thought we could cram a trip to the Keys in to help Cutter. On May 10 we headed down to the Keys, picked up Captain Alma (who would provide local navigational knowledge), and proceeded to search the area for spotted dolphins. We had the public call in reports of dolphins, and we had a plane flying over the area to search, but we couldn't spot any dolphins.

We had a three-day window before the weather got bad, and Cutter was already in his pool in Key West, so everything depended on us finding a group to release him with. On the last day, May 12, and the last hour, we finally found a pod right off Key West. We stayed with this group for hours, following them in a slow westward direction, until the release boat brought Cutter out and we were able to ease him into the water with the group. Satellite tags were just starting to get small enough for dolphins at that point, so instead Cutter had a radio tag on his dorsal fin. Since the tag only had a range of about three miles of signal strength, we lost Cutter's signal that night. But Cutter was given the best chance we could give him to rejoin his group. And sometimes we simply don't know the end of the story.

In 2018, we had a happier ending with the rescue and release of a male spotted dolphin in the Bahamas. What was fascinating was that, during our field season five months after his release, the dolphin, Lamda, kept closely approaching our research team in the water, as if to garner human social attention—in this case, by touching my assistant, who had been in the water during Lamda's release. After being injured in the wild, Lamda had physical contact with humans for three months as he and his trainers worked on his rehabilitation, exercising his tail stock and his body to help him move again. Of course, the goal of rehabilitation is to get an animal back in the wild. To that end, we try to reduce human contact so the animals can become completely wild animals again. I wondered if Lamda was remembering the fondness he had for humans during his rehabilitation. Although, it remains a mystery whether dolphins remember their time with us after a temporary immersion into the human world.

Shifting contexts—visits in the wild and healing in captivity—alter relationships. This is why my colleagues and I work to establish and continue healthy and respectful associations with the dolphins we meet. If you want to observe dolphins underwater and study their behavior, you better know and understand their rules and routines. You must earn their trust. This is essential if you want to understand and show them you are listening.

2

Lessons from Other Animals

There are many animal families and cultures on the planet.

A vervet monkey suddenly makes an alarm call. He has seen a predator and wants to alert his group to take action. In this case the predator is an eagle, so the other vervet monkeys stay low to the ground and shelter under bushes. Just yesterday a different vervet monkey made an alarm call, but that time the call was for a rapidly approaching leopard. Quickly the other vervet monkeys took cover, but that time, instead of hiding on the ground, the monkeys scrambled up a tree, making themselves less of an easy target for the leopard.

In the 1980s primatologist Robert Seyfarth, behaviorist Dorothy Cheney, and Peter Marler shared their discovery of vervet monkeys' predator-specific use of alarm calls.[1] Almost thirty years later, conservation biologist Con Slobodchikoff and colleagues' long-term work with prairie dogs showed another species was also labeling predators, including humans with guns, in their alarm calls.[2] Both these studies helped dispel the idea that animal communication is only an emotional expression (e.g., yelping from physical pain), and scientists identified more examples. Primatologist Frans de Waal showed us that chimpanzees have the ability to communicate aggression through vocalizations and behaviors.[3] Psychologist Donald Owings and colleagues studied ground squirrel communication in great detail and found complex communication.[4] And biologist Malle Carrasco and team reported that mule deer respond to the alarm calls of yellow-bellied marmosets, with whom they share the same coyote predators.[5] It was becoming clear that animal communication signals were far more complicated than scientists had first imagined. The idea that these signals may carry specific meaning for animals, and perhaps even for their neighbors, was a revelation. In the 1970s zoologist Donald Griffin had already begun questioning scientists' approaches and biases toward studying animal communication, and even suggested that attributing animals humanlike thoughts (anthropomorphizing) might help us think of other ways of asking questions and doing science.[6] Griffin suggested that there might be more continuity between animals and humans than we had been supposing.

Dolphins live in an acoustic world and have, over time, evolved a complicated array of sounds to use in different situations. When I am observing a dolphin fight underwater, I hear loud pops and squawks while the dolphins charge and bump each other. The speed of their sounds transmitting in the water rivals the speed of their body movements. Dolphin researchers often use slow motion to review videos, and this enables us to see, for example, five different open mouths within a second. Without this use of technology, we simply miss these details, and we don't know who is communicating. After studying the dolphin's sounds and behavior for decades, my team and I can hear the complexity of sound types and predict the behavior well before we see it. If we, as relatively awkward human observers, can hear these details, I can imagine that the other dolphin groups nearby—who are better suited for the environment—can also hear and decipher these calls.

From detailed communication research on many social species in the wild, we have learned that animals use many different senses to communicate. This includes gestural signals (chimpanzees), color-changing signals (squid), light polarization (bees), and complicated acoustic signals (whales and dolphins). Human hearing is merely a species-specific design, but hearing is not vital to the potential of language comprehension or production in other species. Yet our own acoustic abilities are thought to have an advantage because of sound's ability to travel over long distances and encode large amounts of information. And vocal learning, as noted by biologist Vincent Janik, seems to be uncommon in the animal kingdom, limited to a few types of animals, including marine mammals, and conducive to language development.[7]

Many marine mammal scientists have focused on looking at acoustic signals, both because it is what we know from the human evolution of language and because researchers are often limited by their abilities to see visual signals underwater. But imagine if a species had a visual language that fit all the criteria for language, as well as acoustic signals that help modify the visual language. By understanding that vervet monkeys, prairie dogs, and others have alarm calls that label specific predators, and dolphins have signature whistles that label specific dolphins, we have an intriguing pathway to look for language, or at least language-like structures, in animal communication. Because, as far as we know, to be able to talk about things—specific things or times—you must use something like language.

It is very probable that other animals have both emotional signals and specific labels for things. It's analogous to me saying to my child "come here" in a soft tone versus in a loud, sharp tone. Each conveys a different thing.

In the late 1970s animal behaviorist Eugene Morton wondered if there were any similarities between features in sound, like loudness or pitch, between different animals.[8] He discovered that acoustic signals are consistently adjusted across mammals and birds to communicate the same emotional state. He found that high-pitched quiet squeaks communicated appeasement and fear, and low-frequency, harsh, loud sounds communicated aggression. This "universality" of features really gives us a sense of possible mechanisms that could allow not only for within-species communication but for understanding how various animal species might communicate with another species.

Communication and language can be adjusted by not only loud and soft tones, for example, but also by visual signals. I might add extra emphasis using visual cues; for example, I could open my arms to my child while softly saying "come here," or I could put my hands on my hips while saying it loudly. We call this form of communication "multimodal" since we engage two different sensory systems. Each sensory system can be modulated (loud, soft, long, short) in different ways. But suffice it to say that animals likely have many different signals that modify a message's meaning in many ways.

So, although vocal signals often mediate social interactions (aggression, mating, etc.) in many species, there are other types of signals that exist (chemical, electric), and these can also be modified to express, or hide, information. Evolution in an environment determines each encoding strategy. For example, dolphins evolved in water, so they have a highly complex sound production and hearing system. Dogs evolved on land and are particularly good at using smell. Cephalopods evolved closely with other cephalopods and use large eyes along with color and pattern-changing abilities to communicate.

Awakening the Senses

Animals receive and send information through different and specific media channels (visual, acoustic, etc.). Our human sound perception is limited to specific frequencies, so some animal communication is imperceptible to us. In addition to understanding the limits of our hearing, we must consider how rapidly information can be transmitted and how far it can go. For species that live close to each other, such as fish on a reef, there may be no need to transmit over long distances, so it is possible complex communication has evolved on a smaller spatial scale. What kind of information is necessary to transmit is another question. For an animal that has a short life (<2 years), such as an octopus, there may not be a need for complex relationship information to be available. It may be enough to transmit mating information. For long-lived (>20 years) social species, such as dolphins, or species living in large colonies or societies, such as primates, relationship and historical information may be necessary to form and maintain group cohesion.

The animal kingdom includes a diverse group of beings with a variety of sensory systems. That means we see a variety of signals and communication systems that can be adjusted to encode complex information. Zoologist Arik Kershenbaum and others have described and reviewed many of these systems in the exploration of what's called xenolinguistics—the study of potential nonhuman languages—specifically those that we might find on other planets without the continuity of evolution on Earth.[9] As a prelude to understanding potential life elsewhere in the universe, researchers are assessing the modalities and abilities of animals on Earth. Many different sensory systems are used for communication and are found in a variety of animals, including mammals, insects, cephalopods, and birds. Some senses are older than others, and some have more advantages. For example, acoustic communication is common in animals because it can travel out of sight of an individual and be encoded in a variety of information-bearing ways. Chemical and visual signals can also be complicated but don't carry the advantage of long distance. Depending on the density of the medium (e.g., forest or water) and the signal type, all sorts of information can be transmitted and received. Some signals are of short duration (like the quick sound of a shrimp-claw snapping), while others can linger over significant periods of time (like the chemical markings of a dog). Of course, humans are visual creatures who tend to think of the world as a place to see. But other animals hear, smell, and feel all sorts of cues, and although they may see, it may be in a frequency that humans can't perceive (e.g., ultraviolet or polarized light). Ed Yong, in his recent book *An Immense World*, has meticulously documented and described many animal senses.[10]

The chemical sense is perceived both through taste and smell. Since chemicals diffuse slowly, they are unlikely to contain any urgent information. Although some animals can extract spatial information from chemicals, this requires the right medium (like water or soil) for the signal to flow. We might expect chemical information to be most valuable for short-distance communication or for information that needs to linger (territory), but it may not hold much complex linguistic value.

Magnetic fields have similar properties to electric fields, yet magnetic communication is largely absent in most species. Although humans and some other animals can detect and use small variations in magnetic fields for migration, hunting, and orientation, scientists have yet to see magnetic fields as a means of communication between individuals.

All living creatures use electrical fields (e.g., to transmit signals between organs and tissues), but research has yet to show any species with electrical signal decoding abilities for language. There are a few species of fish well known to sense and generate electrical fields (e.g., African electric fish) and even hunt in groups. Electrical fields can generate information rapidly, thus making them good transmitters of information at short distances. In addition, electric fish use signals creating higher voltages (10 V) in the lower frequencies (25–1,500 Hz), enabling them to encode simple types of information (e.g., sex, species, and social status). However, it is unclear whether any species has evolved decoding organs that might allow reception and deciphering of complex electrical signals, especially at a distance. Like chemical signals, electric signals are close signals (for hunting prey) and might function in small territorial ranges, or small-time limits of a signal, of a species.

That leaves three main systems for complex encoding and decoding: tactile, acoustic, and visual. All signals can be adjusted to encode information. Some signals are close signals (touch), while others can be long-distance signals (sound and vision). Although the tactile sense may require closeness (at least in humans), dolphins can feel sound, which means they can perhaps tickle or ping each other at a distance. That is because sound can go through flesh in water. Essentially sound goes through a dolphin's body until it encounters air or bone. And dolphins can make high-frequency sounds called ultrasonic sounds (meaning above our human hearing range). This gives dolphins a lot of bandwidth to play with to adjust and encode information.

Water is a great transmitter for sound, which travels about four times as fast in water (4,500 ft./sec.) as it does in air (about 1,129 ft./sec.). And although a dolphin has some specialized aspects to its inner ear (e.g., for high frequencies), it is essentially like a human's inner ear. The important point is really that sound is excellent to use for long-distance communication since it travels well through air and water. Vision, on the other hand, will be limited underwater by distance and of course the abilities of the eye itself. An animal may do all sorts of postures, twitches, or rapid movements, but the eye of the receiver must be able to decode and see those details.

Another great example of long-distance transmission of sound is the use of low-frequency, or infrasonic, sound (meaning below 20 Hz) by elephants. Low-frequency sounds have long wavelengths, which means they can travel far. These infrasonic sounds, created in the vocal folds of the larynx, can travel through an elephant's footpad, through the medium of the ground, and be received miles away through the foot of another elephant. So, sound remains one of the best communication systems to use, both for near and far communication.

There are many ways signals can be adjusted and modulated, and such changes take similar forms across senses. Visual signals (like postures) are commonly adjusted and used in the animal kingdom to communicate a variety of information. These signals can be used to communicate many things, like the intensity of a fight or the state of fear of an opponent, for instance. The bee is an example of an insect that adjusts its visual signals to communicate detailed information to hive members. You may have heard of their "waggle" dances, as described by Karl von Frisch and later by biologist Ruth Rosin, where bees adjust their movements, telling other bees the direction and the quality of a food source.[11]

Encoding and Decoding Signals

At this point, we've discussed the ability to send messages only to the intended recipients. But animals can hide messages from their competitors and enemies, including potential predators. Crypticity refers to the hidden information embedded either in a larger sensory context or an environmental pattern. For example, visual crypticity might involve an octopus changing its visual pattern to blend into the environment. This is usually a response to predation. But selective crypticity can be seen in animals when responding to their own kind versus a predator. They can choose to make information available or not.

Cephalopods (squid, cuttlefish, octopuses) have been well studied by marine biologist Roger Hanlon and others, and they are known for their modulation of color, polarization, and timing of visual patterning during various behaviors.[12] They accomplish this by using chemical changes that trigger their light-emitting organs (chromatophores). But animals can also try to hide information from a predator while giving information to their own species. Squid and other cephalopods can use polarized visual signals to send information to their peers while staying camouflaged from predators, suggesting that two separate channels can be used simultaneously, depending on the intent of the signaler.

Zoologist Helen Rößler and colleagues describe how harbor seals may be using a technique to hide from their predators by hiding their acoustic signals in background noise (acoustic crypticity).[13] This suggests a complex awareness of who is around and the seals' relationship with others in the vicinity.

Predators can also hide some of their normal communication signals to fool their prey. Marine mammalogist Lance Barrett-Lennard and colleagues have described in detail how orcas encode their echolocation clicks using acoustic crypticity in the background noise of the ocean when hunting dolphins.[14] Orcas may use high-frequency hunting signals for chasing fish that normally don't hear high frequencies (there are some exceptions), but their cousins the dolphins do hear high-frequency sounds, so the orcas must change their tactics. Biologist Morgan Martin and colleagues have observed that Heaviside's dolphins hide their communication signals from predators but allow their friends to hear these sounds.[15]

Both predators and prey have many strategies of sending communication signals, depending on what is called for. What I find intriguing is that animals can recognize other beings and then decide how to respond. Do they want to be heard and noticed? Do they want to hide the information that they are sharing with their own kind? Do they not want to be discovered at all? This ability suggests that animals are quite aware of who is around and what the dangers, or opportunities, are in every situation.

Could a dog use crypticity (in this case, the hiding of signals within other signals) by hiding a chemical signal within other another chemical signal, realizing that only its friends can understand the hidden signal? There could be other types of crypticity where an organism hides a magnetic signal or an electromagnetic signal. Humans already use electromagnetic modulation to encode information—for example, in the form of TV signals and underwater submarine communication.

Crypticity is by itself a large hurdle to our understanding all animal communication. Without understanding who is around and what their relationship with another being is, it may be difficult for us to decipher any message.

We know, from work by researchers including zoologist Katy Payne, that elephants create infrasonic sounds. Because these are very low sounds they transmit from their vocal apparatus to their feet and to the ground, much like you might feel seismic events. This ability likely gives elephants an advantage to hearing and feeling an earthquake. Journalist Maryann Mott described how elephants seemed to know the 2005 Sri Lankan tsunami was coming.[16] After the elephants screamed and ran to higher ground, the local humans took note and followed them before the first wave hit. Mott reported that dogs would not go outside, and zoo animals rushed into their shelters, suggesting that the animals sensed something dangerous. Because of our human bias, or perhaps our lack of perspective, we didn't even look for low-frequency signals in elephant communication for a very long time. We can understand now why it is important to understand an animal's sensory system, even if it lies outside human perception.

Transmitting Messages

In social mammals, the direction of the message is as important as the information in it. Can you depend on the message? Or is the message a deception? The direction of transmission in most mammals is usually between a mother and her offspring—this is called *vertical transmission of information*. We all know that mother knows best, and this transmission direction is thought to be both stable and conservative, having been tried and tested, and perhaps passed down through generations without changes. A mother's known success for evading a predator usually works and therefore is conserved over time and passed on to her offspring. The negative aspect of vertical transmission is that it can be too inflexible in a time when new information might be needed in a society. If a new predator shows up, a mother might not be able to evade it in the same way she evades other predators, so a new form of evasion may be required. In social species where fathers are known and take roles, such information would also be considered vertical and conservative.

In *horizontal transmission*, information is shared peer to peer. Depending on the age of the individual transmitting, the information can be accurate when sent from adult to adult, as biologist Olga Filatova and colleagues describe for orcas.[17] However, peer-to-peer transmission can be maladapted, such as in juvenile-to-juvenile transmission. One needs only to think of human teenagers, for example, sharing risky information about a certain drug, which can quickly go wrong and cause illness or death. Finally, in *oblique transmission*, an aunt or uncle can serve in an elder capacity to transmit information to an infant or juvenile. Thought to be relatively conservative, oblique transmission can increase the social learning across individuals because it creates many more opportunities for a young animal to get experience, especially in the case of societies that employ alloparenting (babysitting), or where parents can die easily. In my own work, my PhD student Courtney Bender found that Atlantic spotted dolphin mothers taught their calves to forage.[18] Another analysis, by graduate student Gaïane De Brabanter, found that juvenile spotted dolphins did not transmit information horizontally, peer to peer, when foraging.[19] Instead, young dolphins seem to rely on tried-and-true methods of hunting passed down from their elders.

Both personality and societal role can affect how communication is transmitted and observed. In my own work with spotted dolphins, I have observed that females can play an important role in dolphin society even if they do not become mothers. I like to call these dolphins "career females" since they have roles in their society other than motherhood. Why aren't they mothers? It could be that they are unable to reproduce (e.g., to carry a fetus to term). It may be that a diverse range of personalities serves a very clear function in a complex society. Several animal studies—on primates, horses, and octopuses—show how personality and individuality function in a society. My colleague Randy Wells, who has the longest running study of wild dolphins in the world, based in Sarasota, Florida, has documented nonreproducing females in his group and describes a similar function and reproductive status of some of his females.[20] So, while looking at how communication transmits to younger animals, we must also consider the nontraditional roles of aunts and uncles, and how they might affect complex communication signals within a society.

Graded or Referential—the Big Question

A signal is usually either expressing the motivational and emotional state of the sender, or it is labeling something in the environment, or perhaps both. In English, we have words for objects, like *chair* and *table*, and for people, like George. All these words label or refer to something specific (they are referential), and in language, they can be used outside of current time. I can talk about how George built a table for me years ago, or how I bought some new chairs last week. Motivation or emotional expression is thought to be expressed in real time. I am in a heated argument with my spouse, and my voice is getting louder, or softer, to express how I am feeling. For decades most scientists believed that animals only had the ability to communicate their motivation or emotion; these systems are referred to as graded systems, and they communicate the ever-changing intensity or emotion of a situation. I indicate how I am feeling in real time using a graded system of communication. Human language uses both referential labeling (words) and motivation/emotional features (loud voice, whimpers). Emotional gradation can effectively change the meaning of words, or of a communication system, as long as it modifies a referential signal. If I am talking to you and I suddenly start to speak rapidly and get excited, I am telling you not only about the concert I am going to but also how excited I am about going. Human language clearly uses all sorts of combinations of words and emotions. Human bias caused us to assume that animals don't have a lot to talk about, or the intelligence to do so, and this is likely why it has taken us so long to even look for "words" that might be used during animal communication. It should be no surprise that much of the focus on complex communication systems has been with a variety of social species, since they probably have more to talk about in detail. Complexity of social structure and relationships has likely driven the need to communicate complex information.

We need to keep in mind that the complexity we find in a verbal language could potentially be expressed in a visual language. A fascinating example of using both signs and emotion can be found in the use of American Sign Language (ASL). If I sign in ASL, I use my hands and fingers to utter words visually, but I may also put my arms close or far from my body to express other features of emotion while using the same visual words. So, these two features of language (reference and emotional gradation) may not be mutually exclusive in a visual language.

But sound seems to be where we have the largest examples of language, so looking at the complexity of animal vocalizations seems to be a good place to start when searching for language. The most common type of visualization for looking at sounds, both human and animal, is the use of a spectrogram. It's funny—when I was in graduate school in 1983, there were only a few people in the world who could look at a spectrogram and know what words were there. Now, of course, computers recognize our voices and words through other digital processing techniques on a regular basis—for example, when you ask digital assistants like Siri or Alexa to perform a specific task, or transcription features in phones to send a text message. I can ask Alexa what the time is, or what the outside temperature is today, and she will understand me.

Although your phone or digital assistant won't show you this, spectrograms are basically a visual representation of sound, much like a musical score represents notes on a piano. Spectrograms are read like a musical score, with time on the horizontal axis, the pitch or frequency on the vertical axis, and the intensity of the sound represented by the brightness of the signal.

For example, take a sentence from a human exchange (fig. 2.1; spectrogram 1). The variation within this sentence represents an escalating fight between two people. Whiter sound represents louder words, and words that are closer to each other indicate rapid speech. So, words are used, but their rhythm and intensity change over a short period as the fight progresses.

Now take a stream of animal communication sounds (spectrogram 2). We can see that the intensity of these sounds also varies over time as the whiteness, or loudness, increases and the sound spacing changes. But we simply don't know if these sound types

Spectrogram 1

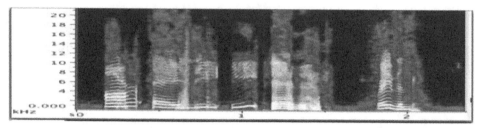

Spectrogram 2

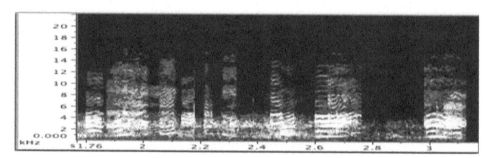

Figure 2.1. Reading a spectrogram is like reading a musical score. Time is on the horizontal axis, frequency or pitch is on the vertical axis, and the intensity of the sound is noted by the brightness. The first spectrogram is human speech. The second spectrogram shows dolphin sounds. Courtesy of the Wild Dolphin Project.

represent words to the animal or just emotional expressions. Normally animal behaviorists would take a video of exactly what the animals are doing—in this case, fighting. And they might note that the animals look more agitated based on their postures while the sounds are getting louder. But how do we know the sounds are words? Do we really know that the animals are just growling and moaning? And thus lies the problem—we assume that animals only have a graded system and that they are not referring to anything. It has taken some creative experiments to figure out what animal sounds might truly mean, beyond just anger or glee.

We have already noted that animals can communicate detailed information, such as predator type, in their alarm calls. Remember that, when an eagle, a leopard, or a snake is present, vervet monkeys produce predator-specific calls alerting the troop to take appropriate action. We have also noted that, like vervet monkeys, prairie dogs encode information for specific predators in their very short alarm calls.

To the newcomer reading a spectrogram, these calls might look, and sound, similar. However, to the experienced eye, the banding patterns in each signal is subtly different, indicating where the information is adjusted to name the predator. Prairie dogs also adjust the number of barks (single barks or repetitive barks) relative to the speed of an approaching predator. Descriptive information about individual predators can be encoded within an alarm call, including the general size, shape, and color of a predator. Different alarm calls were elicited in reaction to the same human wearing a green versus a yellow shirt for five different species of prairie dogs, showing that many different species of prairie dogs can incorporate physical descriptions within their signals. The main point here is that though many signals might look the same, "variants," when checked experimentally, can contain different details and

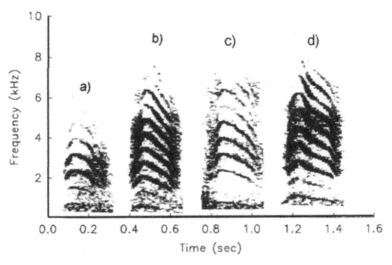

Figure 2.2. This spectrogram represents a variety of alarm calls produced by prairie dogs for four different predators: (*a*) red-tailed hawk, (*b*) human, (*c*) coyote, and (*d*) domestic dog. Only subtle differences separate these sounds acoustically. From C. N. Slobodchikoff, Bianca S. Perla, and Jennifer L. Verdolin, *Prairie Dogs: Communication and Community in an Animal Society* (Cambridge, MA: Harvard University Press, 2009). © 2009 by the president and fellows of Harvard College. Used by permission. All rights reserved.

amounts of information. And it is notable that in both vervet monkeys and prairie dogs, specific labels, indicating the use of words, have evolved in alarm calls. Animal behaviorist Con Slobodchikoff, whose work we discussed earlier, and colleagues suggest that this is simply because, in evolutionary terms, it can mean the difference between life and death to give a specific signal when a predator is around.[21] Imagine I was in a group with you, and I suddenly yelled, "Look out!" You wouldn't know if I was indicating that you should run, take cover, or brace yourself. Now, if I yelled "Asteroid!" you would know that the danger is coming from the sky, and you would look up. Sorry, you might not survive anyway, but it would give you very specific information about what the danger was and where the danger was coming from.

Dolphins also make referential signals via a signature whistle that labels a specific individual—essentially a name. In the 1990s marine biologists Melba and David Caldwell recorded and documented the first evidence of these unique whistles, followed by many other scientists.[22] Since that time, we have discovered signature whistles in multiple dolphin species. Zoologist Arik Kershenbaum and colleagues have identified how various levels of encoding in these whistles might broadcast the identity of an individual and even reference a different individual who may not be present.[23] In the dolphin world, this is an example of an acoustic signal that fits the definition of a referential signal; it labels an individual. There is much promising research on this subject. Animal behaviorists Nicola Quick and Vincent Janik have described how groups of dolphins meeting up at sea may use these whistles to identify themselves.[24] And behavioral biologists Stephanie King and Vincent Janik have described even larger alliances of individual dolphins who retain their signature whistles.[25]

As early as the 1960s, researchers like psychologist John Lilly—a somewhat controversial figure, known among the public for his gathering of researchers (dubbed the Order of the Dolphin) to search for alien communications—and behavioral biologist Jarvis Bastian were already noting that dolphins exchanged sounds back and forth, both in normal behavioral interactions as well as during experimental work.[26] More recent experiments by biologist Holli Eskelinen and colleagues have again shown that dolphin-to-dolphin exchanges do occur—specifically during cooperative tasks, suggesting that there is more cooperation mediated by vocal exchanges that we thought.[27]

Various researchers, including Can Kabadayi and Simone Pika, have described ravens not only using tools but also referential gestures and acoustic calls (yells).[28] Long known for their smarts, corvids in general provide another example of a species that has a complex communication system yet to be illuminated.

Although alarm calls in many species are a good starting point for specific information to reside because of the survival and immediacy factor, it is not inconceivable that animals can communicate a variety of things to each other besides a type of predator and their emotional state during survival challenges. Technical and data-collection challenges have hindered our ability to decode animal sounds in the past; we are only now equipped to handle these challenges.

Names and Dialects

There are lots of types of information encoded in our own human languages. I have a name. That name identifies me as an individual from a specific family and perhaps even heritage or larger clan. I may have a dialect that identifies me as being from a specific region in the United States. (How a stranger can recognize that I am from Minnesota from a brief conversation still amazes me.) And I speak English, which places me in a particular part of the world, in combination with my dialect. And that's only the beginning. As we are conversing, you are hearing words that "refer" to something, but you also might sense my mood or emotion as I speak louder, repeat things, or—combined with my facial expression—give away my anxiety about something.

So, what types of information do animals communicate? It turns out they communicate many of the same things, albeit sometimes through different sensory channels. Let's take dolphins as an example. A dolphin named Brush is from a specific Atlantic spotted dolphin family I work with in the Bahamas. She has her own signature whistle, and her immediate family members (grandmother, siblings) have types of whistles that are similar. In fact, her whistle is much like some of the closest members of her community. But dolphins living in other areas in adjacent groups have different dialects. Marine mammalogist Susan Baron and colleagues have described dialects for dolphins living along the Atlantic coast and different dialects for dolphins along the southeast gulf coast.[29] Biologists Carmen Bazúa-Durán and Whitlow Au have discovered that spinner dolphins living in Hawai'i have separate dialects, and some of these dialects are closer to the dialects of dolphins in French Polynesia than to the dialects of dolphins in the closer Hawaiian Islands.[30] This is all to say that it is important to denote the group you are from, at least in the dolphin world, so dolphins often have dialects. And it might be even more important when you are living in proximity to another group.

Orcas in the Pacific Northwest of the United States have a unique social structure: they stay with their mother in their maternal grouping. Biologists John Ford and H. Dean Fisher have discovered that each of these pods has its own dialect.[31] To mate and reproduce, orcas must occasionally interact with other pods. But what happens when these groups meet up? Can they understand each other? If so, how? It turns out that they have a small subset of vocalizations that they only make when they intermingle with other groups. Evolutionary biologist Laura May-Collado recently described changes in whistle structure during interactions between two dolphin species in Costa Rica.[32] In a sense, these mixed meetings have allowed the creation of signals that can be shared to communicate through a meetup. So, in both these cases, it seems it's more efficient to create a small, shared signal repertoire than it is to learn your neighbor's signals. And it makes sense, considering the animals probably have limited exposure to the other pods, so they don't have a way, or the time, to learn the other's language. Similar to marine mammal researcher Whitney Musser and colleagues' description of captive orcas sharing signals with Pacific white-sided dolphins, biologist Mel Cosentino and colleagues describe a lone common dolphin producing clicks that were similar to those of her neighboring porpoise cousins.[33] So it seems that it is advantageous to at least share some signals with your neighbors to allow peaceful communication.

A few things are clear when it comes to identification. Humans have unique "timbres," or sound qualities, to their voices, as well as names. Dolphins emit unique whistles (a tonal whistle contour) that label individuals, but it is unknown (and researchers have looked for this) whether their other vocalizations carry a quality or timbre to impart identification. Other animals, like dingoes, Mediterranean monk seals, and orcas, may have a quality or timbre to all their vocalizations but not specific name labels.[34] As well, many animals have group identities they can communicate, usually through a dialect. Last, if you are meeting up with another group, you can adopt some agreed upon or mutual signals to communicate in the short order, whether you belong to the same or different species. In some cases, as we will learn in the next chapter, this is easier than learning the complete vocal repertoire of your neighboring species.

If you want to add another layer to understanding complex animal communication, you can begin correlating behavior with vocalizations. And context is everything. In the Bahamas, I had, over the decade, correlated basic sound types with behavior: squawks and screams during fighting, signature whistles during mother and calf reunions, echolocation clicks during foraging behavior, and genital buzzes during courtship activity. It is most important to remember that behavior is contextual. For example, a mother dolphin swimming upside down, or inverted, after a calf is an indication of forthcoming discipline, whereas the same inverted posture in a male dolphin chasing a female indicates an intention to mate. The important thing is to know the players (their age, sex, and history). I observe what transpires during the interaction: the calf is held down as an act of discipline or copulation occurs between male and female dolphins.

Jane Goodall described six chimpanzee courtship gestures observed in various sexual contexts. Anthropologist Barbara King described the complexity of the bonobo's rocking gesture in seven different contexts, including variations in the degree of rocking, with and without arm raising, and so forth.[35] Such subtle variations, potentially ascribing different meanings, may be analogous to the multiple functions we see in the dolphin tail slap (attention, annoyance, emphasis) and inverted chases (discipline, mating). It is the constellation of factors that are recognized and negotiated by individuals in the society that make up a communication system. Communication signals are multimodal and include body postures, types of touch, eye movements, body movements, and angles of orientation; they all play into the process of communication in complex species.

Sequences are also important when looking at animal behavior. Rarely, however, have researchers analyzed the sequence of signals and their changing amplitude and modulations. Is it in the escalation of such signals that meaning is found? If your voice gets increasingly louder during an argument, does this mean something between you and the person you're speaking to? During dolphin fights, the escalation of aggression shows increasing numbers of signals, and the amplitude, or loudness, also increases as the situation reaches a crescendo. Analyzing sequences may be as important, if not more important, than the structure of an isolated whistle or click. Measuring signals, whether sounds or body postures, in order of increasing intensity or subtlety, and observing the subsequent behavior, may be critical.

Even rhythm and intonation may be subject to subtle sequencing during a behavioral interaction. In humans, rhythm and intonation develop before words, suggesting that such a process may be critical while growing up. This may also be the case for primates; for example, the chimpanzee "pant hoot," as described by Adam Clark Arcadi, is uttered with voice intonation and rhythm, suggesting that these calls may be sensitive to the real-time shifts of social interaction.[36]

Another aspect of communication is the audience effect. *Who* is there, potentially watching or interacting, may affect the signals that an individual produces. Is there an estrous, sexually receptive female watching or a dominant male? Is the male a coalition partner or someone who has historically been antagonistic? Relationships can at times be affiliative, antagonistic, or neutral, changing with the audience. It should not be surprising that dolphins, like other animals, may show real-time negotiations. At a distance underwater I can hear screams and whistles, and I know, from my decades of observing aggression, that a group of dolphins is fighting. Likewise, the vocalizing dolphin's ongoing sound and behavior is heard and acted upon by other dolphins, some rushing to the scene, others fleeing. It is rare that a dolphin doesn't have an audience.

Two large issues dominate when trying to understand dolphin acoustic communication. One is the difficulty in determining who is vocalizing, and the other is the accurate recording of dolphin ultrasound. Dolphins produce (and hear) human-audible (up to around 20 kHz) and ultrasonic vocalizations (above human sounds >20 kHz) that are extremely directional, and the higher the frequency, the more directional the sound. Whether the researcher records the harmonics depends on the orientation of the dolphin to the equipment, and of course it depends on the equipment itself. My personal belief is that we have barely scratched the surface of dolphin acoustic signals because historically researchers have not been in control of the orientation of their hydrophones, or (of course) the dolphins, while recording dolphin sounds. At my own research site, probably the best in the world to observe underwater interaction, I can see which dolphin is orienting to my hydrophone, but even with a thirty-five-year database of individuals and relationships, the entirety of these signals is still difficult to obtain. When a dolphin is alone and close to my underwater video/hydrophone, and when there are bubbles coming out of the blowhole, I am sure of who is whistling. In many situations, dolphins are in groups or do not make bubbles when vocalizing.

Dolphins do amazing, and sometimes frustrating (to the researcher), things, including internally directing sound at an angle not indicated by the position of their head—they have two nasal passages (which merge into one blowhole) and can make two sounds at once, and they can make sounds above our hearing range. Most of our underwater video contains groups of dolphins squawking and whistling at the same time. This is analogous to your family talking at the dinner table. Are you talking about the food on the table or what kind of day you had at work? Are you expressing emotions during an argument? In the past we have believed that only humans are capable of talking about yesterday or tomorrow, a hallmark of language. But are we even looking for it in other complex species?

If dolphins can recognize each other's signature whistles, and perhaps have a "voice" or specific vocal quality embedded in all their sounds, then the dolphins potentially know who is fighting whom, how things are escalating within the group, and which of their friends are in the group. Additionally, we don't yet know if, or where, the "individual" voice characteristics of dolphins might reside. In humans, it is our timbre, the resonant qualities of our vocal chamber, that allows our friends to recognize our voice on the phone or in a crowd. Dolphins have signature whistles, thought to be distinct by their fundamental frequency modulation, and it is possible that dolphins also have unique "vocal" qualities that reside both in their echolocation signals and their burst-pulse sounds. But this is relatively unexplored.

Is It Language?

With all these attributes of signals and communication in mind, we come to the most important question: What is language, and do animals have it?

Language entails many features, including symbol use (words), time displacement (past, present tense), combinatorial aspects (re-combining vowels, consonants), and other features yet to be found in their totality in other species. It is the one higher feature of "intelligence" left for the "other species are intelligent" checklist. Tool use, problem-solving, and other marks of intelligence have now been documented in many species. Yet, the possibility of animal language still eludes us. But does it have to?

We know from many animal studies that comprehension of communication signals emerges before production of communication signals. Even human children comprehend words and actions way before they utter words and put them to use. Young vervet monkeys, as reported by Robert Seyfarth and Dorothy Cheney, learn that the silhouette of a flying bird is a potential threat.[37] Over time, and through observation of adults, they eventually learn to identify the silhouettes of specific predators. Our sensory systems are designed to filter information, but it takes time to learn what cues are important and what can be ignored. This is the basic process for any information flow in and out of our bodies and brains. So, in some ways it is not a surprise that humans don't see many of the signals that animals do; these signals may simply not be relevant to the human's survival, but the animal has learned that they are critical. Likewise, other animals may not understand what signals we are attending to in our sensory world. The exception might be our pets, who are exposed to our communication signals constantly.

What Humans Define as Language

Linguists often include the following as key properties of natural language (versus, for example, computer language): time displacement, arbitrary and discrete units, combinatorial features of units, limitless expression of ideas, and recursive syntactic complexity.[38] As well, a species needs to have the ability to learn language and have adequate memory for it; these qualities can be observed in other species through experimental or observational studies. On the other hand, for us to identify whether the properties of language listed above apply to another species, we need to complete a detailed analysis of that species' natural communication system. This sort of analysis does not exist for many species. It is only once we have studied a species' communication system in detail that we would be able to address whether language-like structures exist. If units do recombine to form the equivalent of words, the next step would be to determine what ideas these potential combinations might express.

Through experimental work, we know that dolphins can refer to objects in their environment, they understand abstract concepts, and they can understand artificially created languages, both visual and acoustic. This doesn't mean that dolphins do these things in their normal communication system. We simply haven't looked closely enough into their world to determine this. We need to do a better analysis of natural sounds, categories, structure, grammar, and function in many animal species. Vincent Janik, a researcher studying the long-term dynamics of signature whistle use in dolphins, has suggested that there is "a large potential" for syntactic information in dolphin sounds.[39] Janik is correct in stating that there is great potential for syntactic elements to emerge in dolphin communication. We know a lot about signature whistles after fifty years of research. Janik has described the importance of the shape of signature whistles and their identity information, suggesting that perhaps we just need to use the right tools and a sufficient database to start looking at other levels of complexity in these communication signals. This is indeed where the application of new tools, including machine learning, can help us—not only mine our data to look for larger patterns but also to help us identify order and structure in an acoustic sequence if it is there.

In my own work, I have been collecting and sorting the vocalizations and underwater behavior of dolphins for over three decades. We began playing with emerging neural network techniques in the late 1990s, hoping the computer programs could complement our insights into dolphins' signal structure. A colleague of mine, Volker Deecke, wrote some of the first neural net programs.[40] In addition to having his insight in the field, we were able to use one of Volker's programs to compare signature whistles between individuals in our dolphin community. Yet, at that time, neural net programs still required you to have sufficient samples, say of an individual dolphin's signature (name) whistle to input into the program for comparison. Essentially you still had to train the computer to know what that whistle looked like, then the program would evaluate how that whistle compared to others. This technology was great for the time, and we saw evidence that the shape of a calf's whistle was not clearly the shape of its mother's whistle; instead, the whistle's form was similar to that of community members closest to the calf's mother. In essence, such a community whistle might give a calf access to, and identification with, a larger community when it needed help or company.

It wasn't until 2010, when I began working with computer scientist Thad Starner's group at the Georgia Institute of Technology in Atlanta, that I started to have access to the emerging programs and algorithms that would create the tools I had been envisioning for decades.[41] These tools would not only help us mine large datasets of sounds but also, with some guidance, categorize subtle vocal signals that the dolphins used during communication. In addition, we were developing an underwater computer system to use for playback and experiments with the dolphin community.[42] Over multiple summers, we brought our prototypes of the wearable underwater computer out to the Bahamas to test and retest. It was critical to make sure the equipment worked well and without error before exposing the dolphins to synthesized underwater sounds.

After two years of working with Starner's group on computers and with PhD student Daniel Kohlsdorf on machine-learning programs—computer systems that can improve themselves with the input of new data—things dramatically changed at our study site.[43] The dolphin community had already survived two years of devasting hurricanes and changes in their group structure in 2004 and 2005. And in 2013 another major disruption occurred, and 50 percent of the previously resident community of dolphins vanished from their home area of at least twenty-eight years, as I described in detail in a 2017 paper.[44] After some scientific exploration and field observations, we found all our missing dolphins (about a hundred miles away and in an analogous ecological environment), and we also discovered that there had been a major food-chain crash in their previous home area. Although we had seen subtle signs of changes, like lack of squid and flying fish—a dolphin favorite—in the nearby deepwater edge, it wasn't clear until the dolphins left that the food chain had likely crashed. The crash left scant resources to share, so part of the community simply went elsewhere to take up residence. A similar, but more gradual, displacement of resident bottlenose dolphins occurred in our major study site, from one sandbank, Little Bahama Bank, to another sandbank, Great Bahama Bank. A once bountiful place to collect data on the underwater lives of dolphins became a field site to monitor for environmental changes and disturbances. Previously a healthy and stable wild population, the dolphins, like the rest of the world, were experiencing changes due to human impact and climate change. So, although we kept recording basic life-history information, we adjusted our field methods to monitor this rapidly changing situation. But with a thirty-five-year dataset already in hand, we could still apply these new machine-learning techniques to the dolphin's communication signals over the years.

Powerful Help Is Here

There are emerging studies using deep learning—an even more powerful tool that layers neural network upon neural network, increasing the pattern recognition abilities of these tools from the computer sciences—to help categorize large datasets of animal vocalizations. Workshops sponsored by our scientific institutions regularly explore issues including how to analyze sequences, new algorithm development, and how to best synthesize tools and technologies around the study of animal vocalizations. As we let go of past mistakes and a previous lack of powerful tools, we can apply these new techniques to this field of study.

When I think of the potential for future discoveries, I think about past speculations on planets outside our solar system. Societal paradigms had trained many to assume our solar system and Earth were unique places in the galaxy. After scientists spent decades developing technology, like NASA's Kepler space telescope, to search for planets outside our solar system, we rapidly accumulated real data suggesting that planets are quite normal around most stars—and some may even have characteristics of environments that are conducive to sustaining some sort of life.

Likewise, with the right set of tools and data, we might finally be able to address the long-standing question of the existence of language-like structures in animal communication systems, allowing us to discover a whole new world here on Earth.

3

Eavesdropping on Dolphins

There is nothing more powerful than having another creature assess you.

The first time I recorded a dolphin, water conditions meant I only got a hazy glimpse of a bottlenose dolphin in a head-to-head posture with some spotted dolphins. I remember thinking of the two species' interaction, "Too bad the video will be murky. This seems so unique. I will probably never see this again." But I was wrong. Of all our encounters, my team and I have observed spotted and bottlenose dolphins together 15 percent of the time.[1] Much of this time the two species are traveling, playing, or socializing together. Over 30 percent of the time the males are fighting; often a male bottlenose will bully and mount the young male spotted dolphins. As they are usually outnumbered, it is dangerous for young male spotted dolphins to try to escape, because it leaves the spotted dolphin vulnerable to even more damage from the bottlenose dolphins. My colleague Christine Johnson and I also documented cooperative and social interactions between these two species, including babysitting and pregnant females traveling together. This has been one of the most surprising aspects of my own work in the Bahamas. I gradually realized the nature of these intimate interspecies social relationships.

To complicate things even further, interspecies mating attempts are not uncommon between these two species in the Bahamas. I have often thought that the reason that male bottlenose dolphins dominate male spotted dolphins is to ensure their access to mate with available female spotted dolphins. And occasionally male bottlenose dolphins chase and do succeed in mating with female spotted dolphins, despite the blocks and aggression from the male spotted dolphin alliances guarding their females. One summer, Hook, a female spotted dolphin, was the intense focus of male spotted and bottlenose dolphins—forty-two spotted and nine bottlenose dolphins were in the fray. As they often do for dominance, the male bottlenose dolphins held down and mounted Flash, one of the young male spotted dolphins. But Flash and his peer Dash continued to compete for access to Hook but also competed with Liney and Slice, two other male spotted dolphins. As Flash approached, Hook inverted to solicit mating; she tail slapped Flash and darted hastily away. Throughout this encounter both the male bottlenose dolphins and male spotted dolphins repeatedly attempted to mate with Hook. As we had seen all summer, Hook was a focal female (probably in estrus) getting much attention from the males throughout our field season, but, given all the attention, it seemed that her signals were not species specific.

It takes a lot to learn how to become a dolphin, and eavesdropping is one of the main ways young animals learn from their elders. Marine mammal researchers Mark Xitco and Herb Roitblat describe the specific angle and distance that dolphin calves apparently need to hear the echoes of their mother's sonar clicks after she echolocates on a fish.[2] And, as biologist Robert Magrath and colleagues review, eavesdropping is clearly a technique of listening to a different species to recognize predators and danger.[3] So, these types of passive interspecies interactions (listening to your neighbors) are also common.

We have already seen that active signal creation for communication exchange exists between at least a few different dolphin species and between orca pods. But do animals that engage in interspecific exchange need to be closely related? If they do, what senses and communication signals do they use, and how are signals adjusted to accommodate their neighbors? Are the signals the same or different than intraspecies communication signals? I have often wondered this when watching spotted dolphins fight with bottlenose dolphins at our study site in the Bahamas. One of my graduate students, Cassie Volker, considered this question and compared body postures of both species when they were fighting each other.[4] Sure enough, although some of the postures were the same regardless of which species they fought, the smaller spotted dolphin males appeared to use overt signals when fighting with the larger bottlenose dolphin males. Younger male spotted dolphins would suddenly become passive in the water and float, allowing the bottlenose dolphins to have their way unobstructed by the spotted dolphins. I suspect this is most likely to ensure the spotted dolphins are understood—to make things clearer than they might be with a subtle signal.

Given the different communication systems separating different species, what happens when they try to communicate or simply choose to interact? From a biological and evolutionary perspective, is there enough overlap between species to make it possible to communicate? If so, communicate on what level? Or do differences in our sensory systems and communication signals make it forever impossible to have a chat with another species? Many animal communication systems rely on a small repertoire of individual signals, but animals use them in combinations and in circumstances that are context dependent or "situation bound," as animal behaviorist John Maynard Smith has described.[5] So, considering the challenges of understanding communication signals within a species, how then do we look at interspecies interactions and communication? Clearly the differences between species, and sensory systems, can complicate communication.

Divergence is a natural occurrence, especially when group identity is important in many social species. The reality of physical isolation often allows the emergence of dialects in many species. Such divergence in dialects allows one to recognize their own relations or groups. It has also allowed the incredible diversity of the many cultures in humans and animals alike.

Convergence, on the other hand, is driven by groups having similar pressures or needs, either in the physical or social environment. For example, sharks and dolphins have the same torpedolike physical shape because it is the most efficient form to move through the water rapidly. Similarly, social species, whether they be apes

or dolphins, face similar social pressures, including remembering your friends and enemies, the need for forming cooperative
groups, and the evolution of complex communication skills. Thus,
any universals in communication signals may be the result of similar parallel pressures of communicating the same thing. Most species need to communicate fear, aggression, and neutrality. We saw
this when Eugene Morton documented some universal features of
acoustic communication signals within mammals and birds, suggesting that there is continuity in the evolution of communication
signals.[6] The matrix he presented shows both the pitch of sounds
and the harshness of sounds (fig. 3.1). We can see that symbol in

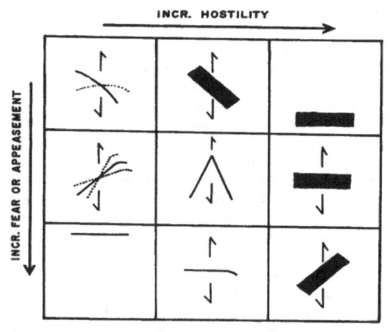

Figure 3.1. Morton documented some universal features of acoustic communication signals in mammals and birds. "Sound structures associated with vary-
ing degrees and combinations of hostile and fearful or appeasing motivational
states. In each block, the frequency is indicated by the figure's height above the
baseline. A harsh sound is indicated by a wide black line, a tonal sound by a thin
line. Arrows indicate the potential for frequency change within each 'motivation square,' and the dotted lines indicate that the figure's slope may change."
Originally appeared in Eugene S. Morton, "On the Occurrence and Significance
of Motivation-Structural Rules in Some Birds and Mammal Sounds." *American
Naturalist* 111, no. 981 (1977): 862.

the upper right represents a very harsh, low-frequency sound, like a growl. On the bottom left, we see a high-frequency, quiet sound, like a whimper. Every other sound is in between, suggesting there is a continuum of information that can be heard as an interaction takes place, either ending in a very aggressive/harsh growl coming from an animal or the opposite whimper of an animal who is fearful.

These prosodic features, like rhythmic vocalizations of howler monkeys or synchronized head movements of birds, may be one mechanism that allows species to understand each other. Necessity is the mother of invention, so the need or necessity to interact with another species may drive the need for understanding across species lines. Now we will explore how other species communicate with each other. We all know about species competition, but what about cooperation and companionship? Are interactions always competition, supporting the survival of the fittest view, or can they include cooperation, suggestive of yet another evolutionary advantage? Recent thoughts on evolution suggest that cooperation in the natural world may be just as important as competition to the evolutionary process.

So how do diverse interspecies friendships adjust for both divergence and convergence? Are there universals that might cross these boundaries between species? Or are the drivers for interaction more social than linguistic? Modern examples of unlikely friendships exist in many different scenarios.[7] There are stories and studies showing interactions from many disparate species of animals, including social interactions between different bird species and different monkey species, as well as interactions between lions and antelope, hippos and tortoises, dolphins and whales, and whales and sea lions. There are also examples of interspecies social interactions in the wild between birds and humans, dolphins and humans, and wolves and humans.

In the laboratory, scientists studying long-lived social animals have often used their closeness with their study subjects to create a working relationship between species. The entire history of work with chimpanzees and other primates includes some sort of interaction with humans, as documented by researchers including Roger Fouts, R. Allen and Beatrix Gardner, Lyn Miles, Penny

Patterson, and others.[8] Scientists who have documented their interactions with marine mammals include Louis Herman, Ken Marten, and Diana Reiss, to name a few.[9] And cognitive psychologist Irene Pepperberg and her African gray parrot, Alex, certainly showed us the true value of interacting with another species as a partner in research.[10]

Learning from Nature

How do different species understand each other in the wild? The most ubiquitous, and likely the easiest, way to learn something from your neighbor is to listen and watch. Since many species evolved rapid ways of signaling danger, for obvious reasons, it stands to reason that understanding short alarm calls, as vervet monkey, prairie dogs, and many species of birds produce, may have a lot of survival value. So, if you are a bird, and you learn the alarm call for a specific predator from a neighboring bird species, it might help you evade that incoming predator. Regardless of whether your neighbor is a bird, monkey, or stampeding herd of gazelles, the ability to recognize other species' alarm signals is a valuable skill. This may be the only thing you want to learn from a neighbor. You may not have the need to communicate back and forth with your neighbor, but at least you can benefit from their watchful skills.

Some of the best-studied examples of understanding between species occur when birds eavesdrop on each other. Many bird species overlap in territories, so they can hear their neighbors during various activities. For example, biologist Robert Magrath and colleagues have described the importance of eavesdropping by young birds, as they are especially vulnerable to predators.[11] Sika deer may even eavesdrop on the calls of macaques to increase their own foraging efficiency, as described by biologist Hiroki Koda.[12]

A variety of monkeys also take advantage of eavesdropping on others to protect their own. Klaus Zuberbühler, who works with eight monkey species in the dense Taï forest of Côte d'Ivoire, frequently observes mixed-species associations, which presumably occur so that the animals can better protect themselves against predation by leopards, chimpanzees, crowned hawk-eagles, and human poachers.[13] Each monkey species has its own species-specific vocal repertoire, and some species possess acoustically different alarm calls for different predators. Monkeys of different species respond to each other's alarm calls in a way that suggests they have a fairly sophisticated understanding of the content associated with the alarm calls.

Not only will some species listen to a neighbor's signal—sometimes they borrow, or co-opt, signals from other species. Mixed-species associations and shared or co-opted signals are noted in the interesting work of behaviorist Branislav Igic and colleagues on deceptive vocal mimicry by birds and of biologist Melanie Ghoul and colleagues on deception between species.[14]

Since most of my research with dolphins involves the observation of dolphins underwater in the wild, and you can't really build an underwater blind to hide from a dolphin, my team and I try to be uninteresting by floating at the surface and hovering, without making sounds or engaging the dolphins by diving down in the water column. Once the dolphins start paying less attention to us and return to their normal behavior, like playing or fighting, we feel we are seeing dolphin behavior without human interference. We also have learned to identify and respect the dolphin's signals, and we try to react to them appropriately. For example, if dolphins display an aggressive signal to us, we get out of the water. We react to their warning signals, especially at night when our human vision is poor and large sharks probably lurk nearby. We have had the experience, when drifting and observing their nighttime feeding, where the dolphin suddenly tail slapped or actually led us back to our dive platform. We take their cue and get out of the water, knowing that they can see things we humans simply cannot at night. And the more frantic and rapid their tail slaps, the more urgent the signal seems to be, so we hurry to get out of the water. On occasion we have verified that indeed there was a large tiger shark right below us. Other times the dolphins simply lead us back to the dive platform on the boat and disappear. Perhaps they do not want to encounter what lurks below themselves.

Despite our efforts, I recognize that learning another species' signals, or entire communication system, may be next to impossible, given the different worlds we live in. So, what happens when you want to interact with another species, but you just don't know their communication system? Is there an alternative to learning the details and complexities of an animal culture or society, since learning another species' signals is both difficult and time consuming? It's probably easier to decide on a few signals you both understand than it is to learn and understand another species' culture in detail, including a language. Historically, anthropologists have had similar challenges when encountering unfamiliar human communities. When communication systems are too complex or foreign to learn, creating a mutual system of signals—a sort of communications bridge—might simply be easier and less time consuming. We have examples in the human world where an "abbreviated" language has expedited intergroup communication to the degree needed.[15] The emergence of Esperanto is an example of creating new languages, and there are many more in human history.[16]

There are a few examples of animals using mutual signals with each other, such as converging whistle structures between bottlenose dolphins and Guyana dolphins in Costa Rica. Here, these two species create and adjust their signals when they are together, but they revert to their own species-specific communication signals when among their own. In Scotland there is a lone common dolphin who produces only porpoise-like clicks when in the presence of porpoises.[17] This raises the question whether such shared signal use is simply an acknowledgment of the other or an attempt to blend in and make a friend. It may be that the need to communicate supersedes any required genetic closeness as a factor that drives mutual understanding between species.

We have already noted the "convergence" of signal structure, or the use of mutual signals, during the mixing of usually separate orca pods in the Pacific Northwest, as described by marine mammalogists John Ford and H. Dean Fisher.[18] During these intermingling events between pods of orcas, the animals clearly share certain calls. Humans often try to mimic local dialects when traveling in areas where people speak the same language; this is a way of saying, "I am like you," or "I am more like you than others."

As discussed earlier, in Cassie Volker's study of aggression between Atlantic spotted dolphins and bottlenose dolphins, male spotted dolphins used a small repertoire of overt physical signals when interacting with the other dolphin species. This could be interpreted as simplifying and clarifying a message because despite another species having a similar body structure and evolutionary history, signals may still need to be made very clear. Not only have my team and I observed these interspecies interactions in our field work visually but we also have recorded vocalizations that occur during these events. How much does one species of dolphin really understand from another? They have similar body shapes and postures, but what about complex vocalizations? If the use of vocalization types evolved within a culture, would their vocalizations be too different, or too difficult, to understand across species boundaries? Even within humans, communication seems to be cultural. Or are there universal languages that transcend these boundaries?

Many animal species appear to have figured out ways to bridge gaps with other species, both passively and actively. So, it seems that other animals are a bit ahead of us. Of course, humans, as the dominant species on the planet, have for the most part felt it unnecessary to learn from, or interact with, other species, with the exceptions of eating them and otherwise using them. Perhaps our pets are the single exception to this attitude.

Signs of Animals Reaching Out

When I was in graduate school in the early 1980s, I would often print spectrograms in my basement. This entailed playing my dolphin sounds repeatedly as the needle burned a carbon image. I would hear the repetitious dolphin "squawk, squawk, squawk" over and over. On one occasion, my Siamese cat, Kashmir, jumped up on the table and watched this process with fascination. Her head cocked with focused attention, she began saying "meow, meow, meow" with the same rhythm and cadence as the dolphin sounds. I guess she saw that my focus was on the machine making the noises, and she sought to garner my attention with her mimicry. It was quite accurate and effective, and it reminded me that animals who are dependent on us often strive to learn and understand how we communicate. I always wondered if this would be true for a curious wild dolphin.

In my fieldwork, my team and I have often observed subtle dolphin body movements during our sessions with wild dolphins. It is normal for juvenile dolphins to mimic what adults are doing around them. Many times, while recording adults fighting and roughhousing, we notice a group of juveniles off to the side doing the same thing. Is this how they learn? Through imitation? During their long childhoods, dolphins are constantly exposed to adult behavior, and they often practice the same behaviors. Practicing a behavior is common to many mammals.

Funny scenes have arisen from dolphins physically mimicking humans at our study site. Humans in the water often try to "dolphin" swim, which means keeping your legs together and undulating from the waist. For the most part, we are bad at it—jerkily swimming while barely moving through the water. One day, as I watched a human in the water trying to do a dolphin kick behind one dolphin, I saw a second dolphin following this spasmodic human while using exaggerated, jerky swimming movements. It was clear the dolphin was imitating the human—perhaps in a mocking way, or perhaps it just wanted to see what it felt like to swim the "human" way.

Another time, colleague and decorated underwater film professional Tom Fitz was on the sandy bottom with his scuba tank, pushing a large underwater video housing in front of him while filming a dolphin digging in the sand for fish. Behind him, a well-known dolphin, Stubby, went down to the bottom and started pushing an empty conch shell in front of him while emitting tiny bubbles (the same size as Tom's scuba bubbles).

We began to explore physical mimicry in different situations with dolphins in the wild. We waggled pieces of seaweed in front of the dolphins, and they would waggle pieces with their fluke (tail) or their pectoral flippers, inviting us into the game. Sometimes I tail slapped at the surface to try to encourage the dolphins to come back to the boat; sometimes they did, and sometimes they didn't (but my sense is that there are many complexities to these signals we are unaware of). Occasionally we humans would lock our elbows while swimming, and the dolphins, after a double take, would position themselves in front of us and rub their pectoral flippers together. There were times I dove down to rub my hands in the sand, and a dolphin would dive down and rub its flippers in the sand.

Sometimes when the dolphins were lying on the bottom, we would dive down to lie on the bottom with them (at least when the bottom was only twenty feet deep). If they rotated their pectoral flipper, we moved our arm and elbow around. If they did a tail-stand on the bottom, we tried to stand with our fins (although, being positively buoyant, I tended to rise quite rapidly). I once took my snorkel out of my mouth and talked in the water so that bubbles would stream from my mouth. Rosemole, a female juvenile dolphin, responded by vocalizing with exaggerated bubbles coming out of her blowhole. Were the dolphins capable of understanding analogues? We were certainly seeing evidence of mimicry naturally, spontaneously, in the wild. Biologist Louis Herman and colleagues have clearly demonstrated in the lab that bottlenose dolphins understand symbolic representations of their body parts.[19] Biologist Mark Xitco, working in the same lab, described the bottlenose dolphin's ability to mimic human behavior.[20]

There is one striking mimicry that illustrates how dolphins can initiate activities with us. In the early years of my research, Paint, one of the young female dolphins, developed her own technique of greeting me in the water. Unlike the other dolphins, who would swim around me with a signature whistle in greeting, Paint would dive vertically in the water and hang there until I went down to hang next to her. I'm not sure how it started, whether I just mimicked her naturally or she me, but that was our unique greeting.

When Paint had her first offspring, Brush, the pair would come by the boat quite often when it was anchored. When I got in the water, Paint immediately dove vertically and hung in the water column with her new calf. As these two surfaced, they stationed in front of me. Then the young calf went down to hang vertically by herself. Was Paint teaching Brush the signal for our greeting? Well, I wasn't going to wait to analyze it. I dove down to hang next to the calf, like I would with her mother, and Brush excitedly swam around squeaking and pec rubbing her mom. This was the Paint family tradition, and I was willing to pass it on if she was. Brush continued this physical mimicking behavior long after she became an adult.

One of the most unusual examples of mimicry involves animals that can imitate other species' specific vocal signals, such as when parrots imitate human speech. This kind of ability for imitation is well documented for several avian species. Conversely, attempts to train nonhuman primates to imitate human speech have been negative, likely due to the lack of adequate anatomical features in nonhuman primates. Even with such different anatomy, the harbor seal named Hoover learned to mimic humans, albeit from a painstaking position for his body, as did an Asian elephant, both of whom we discussed earlier. Michael Hopkin even describes African elephants that imitated the sounds of a truck.[21] And we know that at least some toothed whales (e.g., dolphins, belugas) are good acoustic mimics and sometimes spontaneously mimic human speech.[22] Bottlenose dolphins have also proven skilled at imitating computer-generated frequency modulation patterns, as demonstrated by scientist Doug Richards and colleagues.[23] Different species of dolphins housed together start to imitate each other's sounds. Scientists Elena Panova and Alexandr V. Agafonov describe a beluga whale that started to imitate bottlenose dolphin whistles when the animals were housed together.[24] Examples like this and the one described by biologist Whitney Musser and colleagues, who reported that orcas and bottlenose dolphins shared sounds as tank mates, show that animals reach out to each other in very interesting ways when circumstances require communication.[25]

We have already noted fascinating reports of marine mammals spontaneously imitating human speech, including an account by veterinarian Sam Ridgway and colleagues of a beluga whale that mimicked words it heard from humans who were scuba diving in its tank.[26] And biologist José Abramson and colleagues have reported that killer whales have produced human speech sounds.[27] In my own work, a dolphin named Nassau responded to a human vocalization in the water by exactly mimicking the duration of my "hello" vocalization and the interval between my "hello" and her mimic. As well, during two-way computerized interface work in the Bahamas, we witnessed wild spotted dolphins producing broadband burst-pulse sounds when human speech leaked from a headset in the water. These examples occurred during my many decades of work in the wild.

When a bird, an elephant, or a seal imitates human speech precisely, mimicry is clear. However, animals may make subtle modifications to something we are unable to detect. We may be missing the vast array of animals that are capable of, and are exploring, mimicry in their environment.

Dolphins also use synchrony, or the coordination of body movements, like breathing at the surface together, and they sometimes coordinating their vocalizations. So, what's the point of mimicry and synchrony? In some cases, we see that a species may develop and learn from their elders through the process of mimicry. Biologist Richard Connor and colleagues described Indian Ocean bottlenose male dolphins in Shark Bay, Australia, who use synchronized surfacing to show dolphin cohesion and cooperation to gang up on and chase female dolphins.[28] One of my graduate students, Alyson Myers, described adult male Atlantic spotted dolphins and how they synchronized their physical and vocal signals during interspecies aggression.[29] Synchrony may serve to reinforce bonds needed for different social activities in the natural world.

We know that some animals can mimic not only sounds but also physical behaviors. Are mimicry and synchrony a way in to understanding both intra- and interspecies communication? But when other species imitate humans, what's the point? Is it just to show a human that you are aware of their behavior? Are you trying to get something from a human? Maybe getting attention and some social interaction is the point. And maybe it's enough for intelligent species to spend their time in this activity. Biologist Peter Tyack has noted that the convergence of calls may well reflect the evolution of vocal learning, at least in mammals, and he has discussed its importance in the social world.[30]

This coordinated activity, whether you call this synchrony or entrainment, is one of the most basic features of biology is found on every level of the nervous system and in many sensory systems, as described by researcher Michael Thaut.[31] Marine ecologist Margaret Wilson and psychologist Peter Cook suggest that common rhythmic entrainment aspects in many organisms could be quite widespread in the animal kingdom; this could lead to an understanding of coordination of movement across diverse animals.[32] Even more interesting, and critical to such coordination, may be our abilities to integrate across our senses (like vision and sound), as discussed by scientists Michela Balconi and Maria Elide Vanutelli.[33] In a fascinating experiment, researcher Sivan Kinreich describes how, despite being physically separated from each other, humans brain waves matched up when the participants felt like they were communicating with their partner.[34] What does this tell us? That we have many biological ways, driven through evolution, to encourage certain types of communication. It tells us that there may be great continuity across animal species, including humans. So, aspects of communication, like prosody (rhythm, synchrony, etc.) may be common features across any interactions, within and between species. These aspects could be real and productive ways to proceed in communication studies with other species.

Taking Cues from Animals

As we saw in previous chapters, some very specific interactions happen in the wild that have evolved for mutual benefit. The honeyguide bird and the human hunter both learned that they could benefit from each other's skills. One knows where the hive with honey is but can't get inside. The other can get inside but doesn't know where the hive is. The bird gets the human's attention and, through a series of cues, leads the human to the hive. When the hive comes down, both share in the honeycomb.[35] Even more remarkable is that the local birds recognize the local human whistles, providing a specific interaction that is culturally derived.[36]

It's the same with dolphin-human fishing cooperatives. In these situations, dolphins have learned how to herd fish by taking on specific roles to round up or encircle the school, driving the fish into the fishing nets. Three generations of dolphins and humans have learned to maximize their fish catch by tuning into the signals of the other species. And both dolphins and humans pass on the techniques of interacting to their offspring.[37]

In 2004 the huge Indian Ocean earthquake and tsunami took the lives of approximately 230,000 people. The water surged far inland, causing destruction along the way. Many animals were no doubt killed or displaced. Caretakers Craig and Isabella Hatkoff describe how a frightened young hippo named Owen, separated from his herd during the tsunami and found alone and dehydrated, bonded with an Aldabra tortoise named Mzee at a sanctuary.[38] The 130-year-old tortoise accepted Owen into his world, and they formed a strong bond and lived together in the Haller Park sanctuary in Mombasa, Kenya, as friends.

In another such example, a lioness in Kenya who had recently lost her cub bonded with and protected a young oryx, even forgoing hunting in order to protect the oryx.[39] Suffering from starvation, the lioness was eventually unable to defend her young charge against other lions, but the strange adoption between a predator and her normal prey suggests that all sorts of bonds are possible between species. These social bonds seem to transcend species barriers.

What emerges with these unique observations is that social relationships can cross species boundaries when needed. Whether such extreme partnerships require an understanding of cross-species signals, either simple or complex, we simply don't know. What does one species gain from adopting another? Clearly, they are recognizing a need of sorts. Do they recognize the need for food and shelter in the other species? It appears that, at least sometimes, social companionship may be a driving force and an end in itself.

Hybrids Might Be Normal

Although some species are so genetically different that they cannot produce viable offspring when interbreeding, male bottlenose and female spotted dolphins produce healthy calves. Marine mammal researchers Martine Bérubé and Alex Aguilar report that even blue whales and fin whales have produced fertile offspring.[40] In 1985 researchers Jean-Pierre Sylvestre and Soichi Tasaka reported on various bottlenose dolphin who mated with and produced hybrid offspring with thirteen different species in captivity.[41] In at least one case, where a bottlenose dolphin mated with a false killer whale, the hybrid dolphin was fertile and went on to reproduce. And in the wild, these mobile species evidentially have a relatively easy time crossbreeding, as many hybrids continue to be discovered. For species in the wild, the only way to remain reproductively separated (when the animals are in the same geographic area and are physically able to copulate and reproduce) is by a behavioral reproductive isolation mechanism. This seems to be the main driving force of the male aggression between the two species we study in the Bahamas. During these years of observations of interspecies interaction, I have always suspected that one of the main functions of male spotted dolphin coalitions, and the spotted dolphins' aggressive interactions with male bottlenose dolphins, was to prevent spotted females from interbreeding with the bottlenose dolphins. Hybridization is a natural outcome of such flexible and intimate cetacean social interactions, and my team and I have observed suspected hybrids in our work in the Bahamas, much like the strange body shapes observed by Suchi Psarakos while researching spinner dolphins and pantropical spotted dolphins in Hawai'i.[42]

In the Bahamas, a small subset of bottlenose dolphins, which happen to have black spots on their sides, are friendlier than other bottlenose dolphins. Some researchers have posited that these "spotted" bottlenose dolphins are perhaps hybrids between spotted and bottlenose dolphins, something we hope to look at with their DNA in the future. Spots are thought to function as a camouflage and resemble sunlight streaming through the water and hitting light-colored sand. Indian Ocean bottlenose dolphins do have spots on their ventral side, but if these spots are evolutionarily advantageous, why don't all the bottlenose dolphins in this area of the Bahamas have spots? Hybrid baboons are known to share behavioral and physical traits of both parent species. Researcher Karen Strier and others have described how behavioral traits are shared in the cross-migration and hybridization of hamadryas baboons and olive baboons in East Africa.[43] Could it be that spotted and bottlenose dolphins have been interbreeding in the Bahamas on a regular basis, and that is why they sometimes share morphological and behavioral traits? All dolphins have the same number of chromosomes, making them better able to successfully crossbreed than other animals. They also have great mobility in the open ocean and are therefore better able to encounter each other more than many terrestrial species, which are restricted by mountain ranges and continents.

Could hybridization of spotted dolphins and bottlenose dolphins act as a potential speciation mechanism? But a hybrid dolphin would have to breed with another hybrid to create a hybrid population, or to affect population genetics significantly. In the case of baboons, some hybridization is correlated with habitat loss that forces individuals of one species to invade another species' space. Researchers Sonja Wolters and Klaus Zuberbühler describe two species, the Diana monkey and Campbell's monkey, in two adjacent communities, that have interbred so much over the years that they have produced a hybrid border community.[44]

In the terrestrial world, most interspecies mating encounters are bidirectional, but the physically larger species has the dominant advantage in determining the outcome. Researchers Pamela Willis and colleagues have documented hybridization activity between two porpoise species, the larger Dall's porpoise and the smaller harbor porpoise, in the Puget Sound area near Seattle.[45] Dall's porpoises sometimes mate with harbor porpoises and produce hybrids. These mating encounters are one way: the larger Dall's porpoise males mate with the smaller harbor porpoise females. Therefore, the hybrids have a female harbor porpoise as their mother and a Dall's porpoise as their father. The individual, or species, that is dominant has greater control over the mating outcome, either by pure size or rank. The subordinate has less opportunity to shape the outcome of an encounter.

My work with spotted and bottlenose dolphins in the Bahamas shows a strong tendency toward bottlenose dolphin males mating with spotted dolphin females. Yet female bottlenose dolphins seek out and successfully copulate with male spotted dolphins on occasion, although the males in this case may be sexually immature and incapable of reproducing. My team's only visual observation of a hybrid in the Bahamas comes from our secondary study site in Bimini, where the mother was a spotted dolphin and the father a bottlenose dolphin.[46] Perhaps in the future genetics will provide the information to verify whether hybrids exist in this area. The impact of hybridization is not well understood in cetaceans, but it appears to be a widespread occurrence. Hybridization as a potential speciation mechanism has been overlooked over the centuries, and we may find that in now-contracting ecosystems, animals find compatible mates beyond their own species.

Not only does our new and broadened understanding of interspecies behavioral interaction suggest a dynamic evolutionary social process—it also suggests that interspecies hybridization, or gene flow between species, might be a significant contributor to adaptive evolution. Recently described by biologists Daniela Palmer and Marcus R. Kronforst, the famous Darwin's finches have shown that hybridization, via the interaction between species, could be a real driver in rapid evolution.[47] It may be that the interbreeding of microbial and higher organisms and interspecies interactions all form an intricate and powerful structure for evolutionary continuity, flexibility, and diversity.

Alloparenting, the care of young by a nonparent, has been observed by many researchers and occurs both within and between cetacean species, including sperm whales (genus *Physeter*), spinner dolphins, killer whales, harbor porpoises, beluga whales, and Atlantic white-sided dolphins. In the Bahamas, we have observed young adult female spotted dolphins babysitting bottlenose dolphin calves and engaging in many other mixed-species activities.

Alexandros Frantzis and I described three species—Risso's dolphins, common dolphins, and striped dolphins—that interact in the Gulf of Corinth, Greece.[48] Researcher Arda Tonay and colleagues described interactions between bottlenose dolphins and harbor porpoises.[49] This observation stands out because of the interaction between a dolphin and a porpoise, who are from two taxonomically distinctive families. Dolphins, in family Delphinidae, are thought to be more social than porpoises, in family Phocoenidae. Adoption is also a form of alloparenting, and in Japan, biologist Mai Sakai and colleagues describe a wild Indian Ocean bottlenose dolphin that adopted a neonate (newborn) after its mother died.[50] Cross-genus adoptions of other species of dolphins are also described by biologist Pamela Carzon and colleagues.[51] Although genetic advantages might be obvious when a female helps out another calf in the group, nonrelated adoption suggests that kinship and social relationships in the group do not always play a role in adoption. Biologist Isabelle Brasseur and colleagues reported that a young spinner dolphin was adopted by a resident group of bottlenose dolphins in French Polynesian waters.[52] And off west Iceland, observer Marie-Thérèse Mrusczok and colleagues describe nonpredatory behavior between orcas and pilot whales, including a female orca caring for a long-finned pilot whale calf, even though the calf would normally be easy prey for an orca.[53] There are multiple potential explanations for such interspecies interactions, as discussed by researchers Jonathan Symes and colleagues, and many point the way to social interaction as a driving force.[54]

Does Altruism Cross Species Lines?

In 2017, Robert Pitman, a well-known field observer of marine mammals, and colleagues published a paper titled "Humpback Whales Interfering When Mammal-Eating Killer Whales Attack Other Species: Mobbing Behavior and Interspecific Altruism?"[55] Among the thirty-one detailed observations are specific instances where killer whales were attacking various species, including cetaceans, pinnipeds, and a fish, and humpback whales intervened, resulting in the killer whales' prey escaping. The authors initially explain the harassment of killer whales as mobbing behavior (an antipredator behavior). While there is a kinship benefit to saving a humpback whale calf in your group, there are no apparent benefits to saving another species. Or are there? We know that empathy—action or even observing empathic gestures or kindness—can have health benefits to humans. Could interspecies altruism be more common than we have observed? Or does such altruism only evolve in a species that has itself suffered severe damages and death, or perhaps has had interactions with other species on a regular basis? With the array of video recording devices available today, we can find many videos of unusual interactions of marine mammals in addition to written accounts like one put forth by Pitman. Pitman and his colleagues found that humpback whales either acted defensively or offensively when defending other species from attacking killer whales. Observed to travel miles to intercept killer whales during a prey attack, humpback whales of both sexes contribute to the offensive activities in driving killer whales away. A recent observation in Australia of a southern right whale adopting a young humpback whale calf suggests that large whale species may have a more intricate relationship than previously understood.

Various stories abound of pets alerting their owners to fires, saving babies, and leading humans to troubled animals, such as abandoned kittens. It's easy to understand how a furry family member would be bonded with and care about their human companions. But what is happening with wild animals? Anecdotes of dolphins at sea coming to the aid of both dolphin and nondolphin species have existed for centuries; these stories were summarized by scientists Richard Connor and Kenneth Norris in 1982 and discussed in Hal Whitehead and Luke Rendell's 2014 book, *The Cultural Lives of Whales and Dolphins*.[56] Do dolphins have empathy for the prey of other species? Is it worth interfering at a potentially dangerous cost to themselves?

Net entanglements frequently occur with large whales, and anecdotes, often documented on underwater video, include humpback whales and dolphins calmly allowing divers to free them from nets. Even manta rays have shown extreme tolerance when divers help free them of unwanted trailing fishing line, as shown by researchers Jessica Pate and Andrea Marshall in 2020.[57] Diver Peter Fimrite also gives a striking example of how an individual animal can acknowledge an altruistic intervention or express gratitude.[58] Fimrite describes how a humpback whale, entangled in net, was released by a group of divers. The whale swam in circles and one by one touched or nudged each diver in apparent thanks. In this case, the animal used what it had available to it: touch. This is also how whales and dolphins reconcile with each other. We should expect a species to use what is available to it to express itself in unfamiliar or cross-species encounters.

Could wild animals have a larger "sense of self" or sense of community that we have been previously unaware of? Is that part of the blindness or species biases we have inherited in the modern world? Researchers Ana Pérez-Manrique and Antoni Gomila suggest that animals can engage in perspective-taking that would create this level of awareness and sense of community.[59] Humans have interacted with other species both in the wild and in the laboratory. In both cases, we have learned things about our abilities during interspecies interactions and about individuals of other species. Perhaps we haven't thought hard enough from the animals' perspectives. How might a humpback whale perceive a killer whale? As a predator? How might a humpback whale perceive a human in a boat? In the water? After harpooning the whale's relative? After undoing unwanted fishing line from the whale's tail? As in the case of Lamda, our rescued and released spotted dolphin, and his friendliness with humans after he was back in the wild, a history of interacting with another species can have an everlasting effect on the perceptions by an individual or potentially a community or species.

There is strong evidence for learning across species, and there are many ways an animal can communicate empathy, including the contagious yawning of domestic puppies, as described by researchers Elainie Alenkær Madsen and Tomas Persson, or different types of social learning that animals use, as described by scientist Aurore Avarguès-Weber and her colleagues.[60] In an attempt to look at the evolution of social behavior and intelligence, Maksim Stamenov and Vittorio Gallese have noted that the physical evidence of mirror neurons in many species suggests that many mechanisms for intra- and interspecies empathy have a physical basis.[61]

In my early years of research, there were three main juvenile spotted dolphins that hung out together regularly: Rosemole, Little Gash, and Mugsy. Rosemole was the first to get pregnant, when she was about nine, and Little Gash followed suit. Soon Mugsy was also pregnant, and we anxiously awaited our summer field season to see her new calf. But when we saw Mugsy, she was without a calf, swimming despondently with Rosemole and Little Gash. Both female friends would often sidle up to Mugsy and give her reassuring pectoral fin rubs and other body rubs. Was this empathy? Or sympathy? They were clearly trying to communicate something beyond the usual whistles and clicks. Other species clearly have a large repertoire of signals that can be used in different situations to communicate a multitude of things. And we are only beginning to understand many of them.

4

Talking Back

A species, or an individual, may reach out in unexpected ways—watch for it.

In 1991 I tried my own playback experiments in the Bahamas, hoping that they might illuminate the specific function of the dolphins' signature whistles. I had recorded a few dozen signature whistles (essentially, dolphin names), and I decided to play back the whistle of a young female, Katy, to see what happened. Immediately after playing her whistle in the water, several of Katy's best friends showed up at the boat. Convinced that something interesting was going on, I started playing the signature whistles of newly arrived dolphins. Stubby, one of our regular friendly male dolphins was also present. I played Stubby's whistle in the water. After I played his whistle, Stubby immediately gave me a threatening open-mouth gesture as he swam by me, and then he did the same to the underwater speaker before he swam away. I was horrified at his reaction. Clearly, I didn't know everything about what playing a dolphin's signature whistle meant to the dolphins, or what the proper etiquette was, so I decided to stop playbacks in the water.

Playbacks of sound have been used historically, both on land and in the water, to try to determine the function of a sound, or a sequence of sounds, to an animal. And some playback experiments have been very successful, such as the terrestrial work with vervet monkeys and prairie dogs. But after seeing Stubby's reaction, I decided I needed to learn more about the dolphins' use of their whistles before I tried again. Instead, I would continue to study the dolphins, as primate researchers had for decades, by observing their interactions and behavior during sound production, rather than experiment with their whistles.

In the 1992 book *The Inevitable Bond*, animal behaviorists Hank Davis and Dianne Balfour describe the potential relationships between scientists and the animals they study.[1] The scientist can be perceived in a variety of roles. As a scientist who works in the water, I can say that I really don't want to be considered either predator or prey. I wouldn't mind being insignificant if I could build a "dolphin blind"—the equivalent of the ornithologist's bird blind—in the ocean and just covertly watch dolphins.

I might consider scientists symbiotes, because they provide food and social interactions, and the animals provide the data. But I think our most productive identities as scientists are when we are viewed by the animal as acting as the same species (or conspecific). It may be less stressful for the species under study when a scientist uses signals or physical bodily postures that closely resemble the species at hand. It may also allow some minimal but critical information to be shared between two species. In the wild, human researchers have used the signals of their subjects to further their research. This is clear from early primate work, such as Dian Fossey's knuckle walking with gorillas and Jane Goodall's pretense of nibbling roots with her chimpanzees. In my early work with wild dolphins, I would try to swim beside or ahead of a group of dolphins traveling in the water to get a photograph for identification purposes. I quickly learned that my "place" was behind the group, and the dolphins would always regroup around me and put me in the back once again. But, as discussed in this book's preface, I soon realized I could change the dolphins' direction of travel by using simple head turns (because the dolphins can actually see behind them), something that the dolphins themselves do when traveling. In this case, it enhanced my understanding of who in the group can affect the direction of travel. So, the most productive relationships may require an animal seeing you as a peer, or conspecific, and to be a conspecific you need to be able to communicate directly. There are three factors that have also enhanced interspecies interaction over the millennium: time, technique, and technology.

The Three *T*s—Time, Technique, and Technology

Let's start by considering the first *T*, *time*—the constant and repeated exposure and interaction with another species. Dogs have been around humans for millennia. Prolonged exposure to each other can produce mutual understanding. Recent work by ethologist Ádám Miklósi and colleagues has shown the extent to which dogs can view and interpret the human face.[2] Simply by being exposed to our human communication system over time, dogs have learned to comprehend much more of us than we had thought. Dogs observe our postures and gaze to interpret our behavior. In many cases, they understand human words and can respond appropriately when tested. If you have a dog, you probably know that already. As dogs' dependency on humans for food and shelter evolved, they learned to understand human body postures, gaze, and vocal commands. Some breeds are probably smarter than others, through selective pressure (work dogs, border collies, etc.). Our closest companions have been exposed to and have observed human behavior and communication for centuries, even millennia—a fact largely overlooked until recently. Humans are able to classify dog barks according to different situations, based on subtle acoustic differences, according to work by researcher Péter Pongrácz and colleagues.[3] Because of our close and long-term relationship with canines, adapting species-specific tools for our dog friends has been particularly fruitful. An interesting two-way system is being created at the Georgia Institute of Technology in the Animal-Computer Interaction Lab.[4] In research led by Melody Jackson, dogs have computerized interfaces that register when a dog pulls a cord, or bites a plate, to communicate information back to a human. Named FIDO (Facilitating Interactions for Dogs with Occupations), this project is both practical (service dogs, military dogs, etc.) and a good example of the need for species-specific knowledge when designing an interface. Most importantly, it gives dogs a way to communicate with humans in some detail.

In young human children, comprehension emerges before the production of words. Faced by the same, but more permanent, issue, our companion dogs watch, listen, and respond to our words and emotions with accuracy and use familiar techniques like eye gaze. Some dogs comprehend over a thousand labels or words for toys. They can also report the presence or absence of a toy in a large grouping. This suggests that dogs have quite an understanding of vocabulary. Our pets live with us and are dependent on us, so it is to their advantage to learn our system of communication. What we haven't usually given them is a way to communicate back to us.

The second aspect of communication is *technique*, which involves using new methods for bridging the gap between species. The Irene Pepperberg lab used a technique reestablished from Dietmar Todt's teaching method, called "model/rival" (a bit like sibling rivalry).[5] In the model/rival technique, another species, in this case an African gray parrot, observes the interaction between a human researcher and a second human. In this way, the observing species (parrot) can watch the modeling behavior (by humans) and understand the functional consequences of using new communication signals. It was a way to get the animal in the game, but instead of assuming another species will automatically understand what is going on, the technique gives the other species a way to observe a communication system before it interacts (much like the real world of social development in many species). It appears that, as a result of jealousy, social species can be enticed to interact for social attention, even as applied to interspecies interactions. Although normally used for specific cognitive and two-way communication studies between humans and animals (parrots, dolphins, etc.), it appears that this technique can enhance interspecies interaction.

The third aspect of communication is *technology*, which involves the development of interfaces that are necessary when one species communicates in a different sensory modality than the other. The best example of this type of work is represented by researcher Sue Savage-Rumbaugh's work with a bonobo named Kanzi.[6] Once given a physical keyboard, Kanzi was able to communicate back to humans. A second bonobo, Panbanisha—a peer of Kanzi—showed evidence of creative conversational turn-taking as an additional feature of her interactions, as described by Janni Pedersen and William Fields.[7] No different from humans needing supplementary devices for challenges, animals might surprise us with their knowledge and interest in communicating back. But it took scientists a long time to reach this point in methodology.

A real two-way interface with technology, as in my own current work with dolphins, may offer many advantages, including the ability to build a bridge across sensory systems that may be necessary for communication between species. Storing computerized information to acquire data and process information quickly is also a desirable feature. And, finally, by empowering an animal operator of the system to elucidate previously unexplored cognitive processes, new gains in the field can be made.

Parrots are well-known mimics of sounds and even human words. No technology was necessary to translate when researchers were working with Alex. But the sensory world of most other species (great apes, dolphins, etc.) and their ability to mimic signals humans produce are limited. However, given the tools to communicate back, we might be surprised how easily they might learn, or what they have already comprehended from interactions, and what they know about their world. Most interfaces entail the use of "symbols," the key to language. Such symbols may be arbitrary visual shapes, acoustic signals, or some other sense necessary to one species or the other. Of course, technological interfaces are used to help some humans read, hear, and see in a variety of ways. So, too, do technological interfaces have the ability to accommodate the differences between species. And focusing on a species-appropriate system will be critical to bridge any gap.

Using Interspecies Interaction to Further Science

The use of interspecies communication as a tool for science has a long and complex history. It is worthwhile reviewing and understanding this history, both for the mistakes that were made and for the insights that were gained.

The famous animal behaviorist Konrad Lorenz spent many years studying geese and trying to understand and mimic them.[8] Researchers like Jane Goodall approached a state of acceptance by wild chimpanzees without any attempt at physical disguise. Dian Fossey also attained acceptance by troops of mountain gorillas. These pioneering animal behaviorists opened an enormously powerful new science of "participatory research" in interspecies communication with their discoveries of animal societies in the wild.

Although interacting with animals can allow the scientist to gain detailed insight about an animal society, as we have discussed, it has been necessary to sometimes create interfaces, such as keyboards, to interact. And most interfaces have been created for work with social mammals, specifically primates, and social species of birds. Computerized symbol interfaces, based on finger-operated touch screens, were eventually designed for a variety of animals. We will review some of the major points of departure from the norm and any commonalities we can see from all these methods.

Great Ape Research

In the 1950s researchers Keith and Catherine Hayes tried to encourage a chimpanzee to produce English words using a home-rearing situation.[9] The unrealistic expectation of asking a member of another species, with the wrong anatomy, to produce human vowels and consonants was a major deterrent to success in this work. This work serves as an early example of an inappropriate request to make of another species. In the history of work with captive primates, there have been a continuum of methodologies that researchers have used, from very objective and strict to more interactive, involving bonding with the animals under study. One of the first studies was on a chimpanzee named Sarah, who was acquired from the wild and tested by scientist David Premack.[10] She was trained with the use of a board, testing her ability to discriminate "same and different" objects. Premack remained separate and "objective" during his work with Sarah. His work could be considered on the stricter end of the approaches.

Later, researcher Herbert Terrace tried to test whether chimpanzees could understand and create sentences through the use of

American Sign Language.[11] Terrace raised a young chimp named Nim in a human home and put him through strict classroom training, where he did learn to sign. Terrace's conclusion was that Nim could use American Sign Language with the appropriate cue from his teachers, but he did not functionally understand the order of words. Researchers R. Allen and Beatrice Gardner had already discovered that the more natural communication modality of chimpanzees lay in their gestural articulations. The researchers used this discovery to further engage in sign-language studies with chimpanzee subjects without the strict requirements of past studies to create meaningful sentences.[12]

At about this time, researchers were being questioned about the possibility that their chimpanzee subjects were cleverly taking cues from other signals from their human trainers. Oskar Pfungst, in the 1900s, studied a horse named Clever Hans that was advertised as able to "count," among other skills, by his trainer. Researcher John Prescott later warned that such accidental cues could be detrimental to dolphin research.[13] Prescott noted that the trainer was giving Hans subtle cues, everything from face nods to slight movements. Without these cues, Hans could not perform properly. Thus, a technique called the "double blind" method, was created. This method was based on neither trainer nor experimenter knowing the answer and was created and used to avoid the exchange of accidental or intentional subtle cues from trainer to animal. But what does the "Clever Hans" phenomenon suggest about biological communication in general? It might suggest that these cues are the information that help create complex communication in the first place. If so, then we are asking animals to do what we cannot—to be restricted to one sensory system when trying to communicate. For example, if you put me in a dark closet and talk to me, and then expect me to understand exactly what you meant without seeing your eyes, your hands, and other subtle cues, I might not perform well. It's for that reason that text-message emojis were developed, as they represent emotions or facial expressions that cannot be communicated through text alone. In fact, this realization was the start of using more interactive and functionally expressive methodology for interspecies communication.

Roger Fouts, known for his research with Washoe the chimpanzee, continued sign-language work based on the history of Herbert

Terrace's and the Gardners' research.[14] Fouts reasoned that chimpanzees would be much more interested in working with humans if they liked them and found them interesting. This would provide the social motivation for expressing themselves about something they desired or wanted to express. His study would later document the spread of sign language from chimp to chimp, without human intervention, as an example of cultural transmission within the chimp society itself. Fouts also tried to provide the chimpanzees a rich, natural environment of relationship and communication and used a species-appropriate sense (in this case ASL gestural signs) that could enhance interspecies communication. By watching the chimpanzees sign to each other, researchers were able to observe the transmission of shared gestural information between the chimpanzees, including what chimpanzees were passing down to their offspring. A similar interactive approach and rapport development was used by researcher Lyn Miles with Chantek the orangutan, thus showing the broad applications of developing rapport with another species in the scientific process.[15]

Researchers Duane Rumbaugh and Timothy Gill had also been exploring the benefits of creating keyboard interfaces to use with chimpanzees.[16] This included a touch keyboard that used visual symbols called lexigrams and provided direct treats to a chimpanzee named Lana. A key motivation for using lexigrams was that, as signals, they were less ambiguous than many of the apes' gestures. And it also took humans out of the equation of interacting since Lana interacted only with the keyboard as a means of receiving food. Lana learned to push a sequence of buttons on the computer that would result in sentences, such as "give Lana apple." Although some would argue that Lana simply learned to discriminate between simple associations, it seemed that she was able to make novel connections on her keyboard.

Over the years, Sue Savage-Rumbaugh and Duane Rumbaugh moved back to a more interactive approach with two other chimpanzees, Sherman and Austin.[17] They spent time developing a rapport with Sherman and Austin, and they also tried to develop experiments that included referents or objects that were important to the chimpanzees. Allowing Sherman and Austin to work as a team, possibly mimicking the way communication systems evolved under natural conditions, was very successful. Tests that employed cooperative tasks—in which the animals used the keyboard to make requests for one another's help—were particularly successful with Sherman and Austin. In this case, Sherman and Austin did show immediate transfer of the meanings of their lexigrams on the keyboard to novel contexts, including spontaneous use of signs, suggesting that the two-way, communicative nature of the task facilitated such understanding.

As Sue Savage-Rumbaugh and Duane Rumbaugh began working with Kanzi, the bonobo chimpanzee, the use of a portable keyboard in a seminatural environment seemed to further enhance complex word associations.[18] Although the infant Kanzi was only observing his mother exploring the keyboard initially, he was able to use the keyboard himself without individual training. After that, Kanzi became the main subject of the keyboard work. This portable keyboard is the best example of a tool for bridging the interspecies gap, and humans were now seen as an important social participant in the process of exploring interspecies communication.

Researcher Francine Patterson used sign language to communicate with the gorilla Koko.[19] Patterson used sign language to teach words and ideas to Koko. One of Patterson's most interesting observations was Koko combining the words "sweet" and "water" to describe watermelon. This may show Koko's ingenuity with her restricted vocabulary to recombine words to describe a new object in her environment. However, critics say that because of Koko's close and intimate interaction with a human, Patterson's interpretations are skewed and not objective. Patterson's lack of peer-reviewed scientific publications also hindered acceptance of her interpretations. Patterson argues that her degree of bonding gives her insight into the complexities and nuances of Koko's behavior in a way that strict experimental protocols would not allow.

Despite all the different methods used to study primates, from a strict approach of isolating them from humans with only computers to work with to interacting with humans on an intimate basis, most researchers nowadays agree that creating rapport with an animal creates motivation to engage. A keyboard is boring by itself. Social interaction is rarely boring. The recognition of the social needs of many nonhuman animals has paved the way for new research. And because of scientists' focus on primates, many dolphin studies using interfaces were modeled after primate studies, even though one might argue that they use quite different communication signals and should have their own species-specific interfaces. And that has been the challenge of interactive work with dolphins over the decades.

Dolphin Research

After primates, dolphins probably represented the best candidates for complex communication. Dolphins have the brain structure, memory abilities, and problem-solving skills for such work. Dolphin studies have shown an ability to comprehend a computer-generated sound-based language system with one dolphin and a human-produced gestural language system with a second dolphin. Psychologist Louis Herman documented that dolphins can understand both the order (syntactic) and meaning (semantics) of words when presented with language that has been artificially created (by humans).[20] In the wild, dolphins have long-term, complicated social relationships, including politics and complex signals. Pursuing a language interface seems like a logical next step, since dolphins and other species either use, or can learn to use, referential signals.

The history of trying to communicate with dolphins is quite long and involved. As early as the 1960s, the US Navy, along with researcher Wayne Batteau and colleagues in Hawai'i, devised an acoustic communication interface for humans and dolphins.[21] Advanced for its time, the system arbitrarily matched a dolphin whistle contour with a human vowel or phoneme. A computer translated combinations of human vowels to generate sinusoidal dolphin whistles. These sounds were projected into a lagoon where two bottlenose dolphins, Puka and Maui, learned to successfully execute the commands. The dolphins learned to respond to thirty-five command strings generated by humans using five-word sentence structures. The dolphins also learned to mimic and repeat whistles on command. Essentially the researchers were using an acoustic command system, and although the dolphins could mimic the whistles and learn to understand them as streams of commands, no functional or combinatorial understanding was evident.

Although this study was a species-appropriate signal design in that it was acoustic, the system was only one-way and was designed to give commands to the dolphins.

Around this same time John Lilly began working with dolphins. Eventually, Lilly chose to build a house in the Virgin Islands where a trainer, Margaret Howe, could live and interact with dolphins.[22] Peter, a male adult bottlenose dolphin, was Howe's main subject. In a method reminiscent of Catherine Hayes's early work attempting to shape the mouth of chimpanzees to speak English words, Howe attempted to get the dolphin to mimic English words. Regardless of the species-inappropriate assumptions, some indications that prosodic features (rhythm, intensity) were mimicked are evident on the videos of this work. But no functional understanding of words was observed in this first attempt in a two-way study.

Years later, Lilly started project JANUS (Joint Analogy Numerical Understanding System) at Marine World/Africa USA in Northern California. The interface used human words that were arbitrarily turned into series of high-pitched clicks, around 32 kHz, well above human hearing. The dolphins learned sequences of "clicks" as commands and executed these commands. What appeared to be a vocabulary of forty whistle-words was again a command system that the dolphins learned; the project simply did not show any evidence of functional understanding. As in Batteau's work, the dolphins showed mimicry and an understanding of trained commands. However, the work was not scientifically well documented enough to merit further scrutiny.

Although these acoustic interfaces showed that dolphins had the potential to mimic signals back to humans and to learn to act out commands, these systems were not designed to be two-way in nature. Neither project involved rigorous testing; computers were just emerging on the scene, and few results were published in scientific sources. Only recently has computer technology emerged with fast processing and pattern recognition programs that might have enhanced these systems.

Meanwhile, researcher Louis Herman and his team at the Kewalo Basin Marine Mammal Lab in Honolulu, Hawai'i, were showing us that dolphins could comprehend at least artificial gestural and acoustic languages.[23] Although their research was not focused on a two-way system, insights into the dolphins' cognitive and language abilities were emerging from this lab.

Working at the same facility as John Lilly, researchers Diana Reiss and Brenda McCowan implemented a tank-side keyboard that incorporated visual and acoustic symbols and that was activated when the dolphins, or humans, pushed a physical key.[24] The dolphins began to explore what happened when they pushed the keys on the keyboard. Once the dolphins began to experience receiving high-value objects (e.g., toys) or desirable actions (e.g., receiving a rub) after the activation of a key, the dolphins spontaneously produced mimics (defined as a facsimile immediately following computer production) of the symbol-associated sounds. They also produced facsimiles of these sounds, and combinatorial whistle facsimiles, in behaviorally appropriate play with the objects, and even preceding a key-hit on the keyboard. Sometimes the dolphins spontaneously produced the whistles with each other during non-testing playtimes.

Reminiscent of sign language sharing among the chimpanzees, these were tantalizing results. However, the 1980s was the age of fiber-optic networks, and computers still lacked real-time sound acquisition and analysis. However, this work set the stage for improved techniques and thoughts about "if" you had the technology, what you might accomplish with these animals.

Later, researchers Ken Marten and Fabienne Delfour implemented a computer touch screen with visual and acoustic symbols at Sea Life Park on Oahu, Hawai'i.[25] Since the park did not have poolside or in-water access for the researchers, Marten and Delfour came up with a creative and technically advanced solution. They would let the dolphins activate a computer screen through an underwater window by breaking an infrared beam grid set up in the tank and calibrated to match the computer screen activity inside the lab. Researchers observed through an underwater viewing lab while the dolphins operated, or "touched," the screen by breaking this infrared beam. The disruption of this underwater infrared beam essentially allowed the dolphins to point to the computer screen. Dolphins could create action on the screen (movements, sounds) by pointing to objects and images using the infrared trigger. Humans sometimes demonstrated the contingencies of the system, at the underwater window, while the dolphins watched from in the tank. Much like Reiss's system, the interface allowed the dolphins to do a certain amount of "exploration" of the system without rigid protocols. And researchers interacted with the dolphins either at the surface of a tank or through the underwater window. Both systems successfully incorporated human interaction to encourage attention to the keyboard and to demonstrate associations or contingencies of the communication system.

Once, when I was visiting Ken and Fabienne in Hawai'i, I got involved in playing a game with the dolphins using the computer. I was just starting to think about how one might do this in the wild. I had been working in the Bahamas for almost ten years at the time, and I had seen the willingness of the dolphins to interact with us, occasionally mimicking us both physically and acoustically. My situation in the wild was ripe for an interface, and I immediately knew I wanted Fabienne on my team. A few years later, I paid a visit to the Kewalo Basin Marine Mammal Lab in Honolulu and met up with researchers Adam Pack and Louis Herman, pioneers in dolphin cognition. I explained my unique situation in the Bahamas, with its elements of regular dolphins and their curiosity toward humans. We jumped on a partnership. With their expertise and my situation, together we might create a two-way interface for communication. But we were not the only ones thinking about new underwater interfaces.

Former students of Louis Herman's, Mark Xitco and John Gory, along with dolphin researcher Stan Kuczaj, were building an underwater keyboard at the EPCOT Center in Orlando, Florida.[26] Sponsored by the Walt Disney World Company and placed in a reef setting where dolphins, sharks, and turtles roamed in view of the public, this underwater keyboard was the first to allow humans to work with dolphins underwater using a computerized system. And the dolphins seemed to have plenty to talk about. The keyboard held objects in recessed tubes. A dolphin or a diver could operate the keyboard by breaking an infrared beam in the opening of the tube to "point" to the object at the bottom of the tube. English words were then generated for the diver's knowledge. These researchers incorporated the model/rival technique in which a dolphin watched as a model (human diver) activated a key while another diver responded by going to a location and picking up whatever object was requested through the keyboard.

Like Pepperberg and Savage-Rumbaugh, these researchers used a human-based modeling procedure in which a dolphin watched as a human model and human receiver activated and responded appropriately to key presses. And for the first time, this activity took place underwater for both humans and dolphins. Eventually, each of two dolphin observers learned to respond to human-activated keys and to activate keys themselves, often swimming ahead of the human to the object or location associated with that key.

During this time in the 1990s, my team and I began implementing an underwater keyboard in the Bahamas with our dolphins.[27] You simply cannot tightly control experimental work in the wild with free-ranging animals. The animals move around, leave, or decide not to work, and this makes repeated experiments difficult. You might not have the same individuals around to work with. And since we don't feed wild animals, you don't have the typical "rewards" offered for a correct answer. All you really have is the social reward of interacting. Before our work, I knew of only one attempt, in the mid-1980s with wild dolphins, which had shown promise for two-way work in the wild. A young student named Jody Solo, working with the late Ken Norris (father of dolphin biology) and his group in Hawai'i, attempted to swim out to meet wild spinner dolphins using familiar greeting calls and nonaggressive behavior to gain the dolphins' trust. Although no extensive information came from this attempt, Norris believed that had Solo had an underwater video camera, she might have returned with an intimate glimpse at the life of spinner dolphins in the wild. Norris attributed Solo's acceptance into a wild school to her ability to avoid threatening behavior and to reach out with her natural sensitivities to the dolphins.

When I decided to ask my colleagues, in 1997, what they thought of me trying two-way work using a technological interface, between humans and dolphins, in the wild with my study group of dolphins in the Bahamas, I was pleasantly surprised. Apart from one of my colleagues, most thought it would be interesting and possible, given the dolphins' friendly nature and my decade with this specific community of familiar individuals. The fact that we could observe their behavior underwater was already an incredible opportunity, and I had been documenting these details for almost ten years already. I knew the dolphins and their behavior well. When I visited Lou Herman and Adam Pack in Hawai'i, they were so interested that Adam came out our first summer to help. Armed with solid knowledge of the dolphins and encouragement from peers, I began.

Our etiquette with the dolphins was laid out early. We would not disturb their natural behavior, but if there were interactive encounters, we would engage the dolphins. If the dolphins were foraging or fighting, for example, we would stick with benign observations. If we entered the water and they initiated interest or a game with us, we would move into two-way work. Our human team was prepared for this plan, and with a "start" and "stop" signal on my wristband, I could alert the team. If, during the two-way work, the dolphins began foraging, mating, or other normal activities, we would end the two-way work out of respect for their needs. Since I was in charge of initiating signals with my keypad, I would make these decisions, based on my decade of knowledge about the dolphins' behavior. I knew that there were three key elements involved in past successful interspecies projects. First, humans interact with each other and model a communication system with the animal observing, rather than testing or demanding that an animal suddenly understands what is expected. Second, social interaction between the participants is critical and even more motivating than food rewards. Third, the use of an appropriate communication modality (e.g., visual, acoustic) is essential to maintaining interspecies exchange.

I never thought extensively about our etiquette with the dolphins as an actual "agreement" until I read Jack Turner's dark, but honest, book, *The Abstract Wild*.[28] Turner gives a multitude of examples of covenants and agreements, albeit nonverbal, with other species: an arrangement of place and space. I've always looked at my initial time in the Bahamas as an investment in a relationship: respect and trust get you in, and harassment gets you out. Anthropologists know this: they observe, they watch, they wait; if lucky, they get incorporated into the internal rituals, community secrets, and empowered places not available to everyone. Using our obser-

Figure 4.1. Dr. Herzing (*right*) and Dr. Delfour (*middle*) demonstrate the two-way system by asking each other for the scarf, a favorite dolphin toy, as two juvenile dolphins watch the exchange. Photograph courtesy of the Wild Dolphin Project.

vational methods and knowledge of individual dolphins and their personalities or proclivity to play, we had already been able to establish some ways in which they try to mimic human-synthesized whistles. As a result of us working carefully with the dolphins in the water, understanding their etiquette and behavior, the dolphins began to mimic some of our body postures and vocalizations.

Our technology at the time was rudimentary. There were no mobile underwater keyboards in existence. So, we designed one: a small aluminum bar with four small underwater boxes attached to it. Each box contained a portable CD player that had one prerecorded synthesized whistle on it, and each box had one visual symbol on the outside: a star, a moon, a cross, or an infinity sign. We had both visual and acoustic symbols incorporated into the system. The boxes were mounted on the aluminum bar so we could push the bar through the water as the dolphins traveled, and we looked for an opportunity to use it.

Figure 4.2. Dr. Herzing (*left*) points to the visual symbol for scarf on the keyboard while offering the scarf to the dolphins. Photograph courtesy of the Wild Dolphin Project.

When the dolphins became playful or interactive, we would demonstrate the use of the keyboard. I would go to the keyboard and point to the symbol representing a toy. The operator pushing the keyboard would play the sound on the box and then hand me the toy. I then swam happily off with my toy and engaged my fellow researchers, or dolphins, in a game. If anything, animals need to be motivated to use a communication system, and play is a great motivator for dolphins. If a colleague in the water wanted my toy, they would go over to the keyboard and point to the same symbol. I would hear the sound (if I wasn't looking at the keyboard), and I knew that someone (person or dolphin) at the keyboard had requested my toy. Being a "good" dolphin, I then would swim toward the keyboard and offer them the toy. The idea was to let the dolphins use the keyboard in the same way. More than anything, researchers have noted that it is important to demonstrate a keyboard system to an animal to help them understand what can be gained from using the keyboard. So, time and time again, cognitive biologist Adam Pack, or cognitive biologist Fabienne Delfour, and I would use the keyboard in front of the dolphins. Most of the time the dolphins would circle us and just observe. They really were watching. The challenge with our technology was that it was unable to let us know if a dolphin that was at a distance from the keyboard made a whistle requesting an object. Remember these are wild dolphins, and they are not tank-side pressing a keyboard. They are swimming in the ocean around us—and, being acoustic mimickers, they might just choose the acoustic path. However, during many encounters we had, we were able to lead the dolphins to the physical keyboard and let them observe the box and hear the sound. At the end of the day, we were able to look at our underwater video and sound and see what potential mimics occurred during these sessions, but it was hard to react in real time to any responses the dolphins might have made.

Now we had experience with underwater systems where humans and dolphins worked together. In these situations, humans could demonstrate how the system worked and what the system got you if you used it correctly—both important points. A negative aspect of working in the water as humans is that we are very slow compared to dolphins, so it's hard to keep up with the dolphins if you are trying to move to a specific location. We saw this at the EPCOT Center when a dolphin would request that a diver meet them at the reef (about thirty feet away), but by the time the human swam over to the reef, the dolphin had been back and forth to the keyboard and simply got frustrated.

After 2004 I decided to forgo our two-way keyboard work in hopes of finding some better technology to use underwater in the future. In 2004 and 2005, we had major hurricanes go right over our study site, decimating the dolphin population. In 2010, after years of watching the dolphins recover from their losses and stabilize their society, we were ready with some new technology to try again.

When primate researchers demonstrated that chimpanzees could see themselves in a mirror, dolphin researchers jumped at the opportunity to try this same test of awareness on dolphins. But you might be wondering why a dolphin researcher would choose to use a primate method, mostly visual, on dolphins, who are acoustic

Figure 4.3. A diver works with a dolphin at the underwater keyboard at EPCOT Center in Orlando, Florida. Each participant can activate a word by pointing inside the keyholes. Photograph courtesy of Mark Xitco and the Walt Disney World Company.

creatures. Why did we follow a primate analogy for our keyboards versus developing species-specific interfaces for dolphins? Researchers probably followed these techniques because they were accepted and because, being terrestrial creatures ourselves, it is simply easier for us to operate something visual and tank-side than it is to operate an apparatus underwater and acoustically. Over the decades, dolphin researchers had used a variety of triggers, including "physical touch" activation, "infrared beams," and "acoustic triggers." What the field really needed was a dolphin-friendly acoustically activated system. Dolphins may be more comfortable communicating using whistles and squawks as compared to pressing a physical keyboard (designed for primates).

Just lurking around the corner was the acoustic trigger system we needed. Dolphins naturally point or "buzz" with their echolocation clicks on objects when they first approach something. Researchers Mats Amundin and Josefin Starkhammar were developing such an acoustic tool that the dolphins could use to explore, or trigger, naturally using sound.[29] Originally designed to analyze the scanning sonar of dolphins, ELVIS (Echolocation Visualization Interface System) could, in effect, act as an interactive acoustic "touch screen." The system also could project visual symbols on the tracking screen to associate sounds with a symbol, creating a high-tech keyboard. This system had great potential for both cognitive and communication studies, but it hadn't been used in other scenarios.

Since it wasn't practical to have ELVIS in the water while swimming with wild dolphins, I opted for another option. A true two-way interface for animals and humans would empower the animals to communicate their desires to us, rather than simply receive commands from us. I wanted to wear a computer (not umbilically connected to the boat) that the dolphins could trigger with their sound, that could do real-time sound analysis and translation, and that I could swim reasonably well with in the water. It turned out that our newly adopted computer scientist, Thad Starner, was the technical lead on Google Glass, a wearable computer that displayed information on a pair of glasses. In 2010, one of Thad's students at the Georgia Institute of Technology had emailed me asking if we had any unprocessed datasets that we would like help analyzing for patterns. Of course, I had many acoustic files that I would have loved to have analyzed, but I also needed help with our underwater interface. A few months later, Thad showed up at my laboratory wearing his futuristic-looking computer. After showing Thad some data and discussing our hopes for categorizing sound types, I told him about our underwater keyboard work. Thad immediately offered to help us build an underwater wearable computer. Named CHAT (Cetacean Hearing Augmentation Telemetry), the computer allows divers to exchange requests with each other while under-water.[30] The computer was trained to recognize, in real time, any sounds made by the dolphins that matched our labels for objects.

So, by 2013, we were ramping up our underwater two-way com-puter system, and it progressed well over a few years. The alumi-num boxes that housed all the hardware and software were built. The keypad for the researcher to trigger sounds and the underwa-ter speaker to play sounds were built, and bone-conducting head-phones were enlisted to allow the researcher to hear, underwater, what the computer was doing.

Then there was the issue of software, because we wanted real-time sound recognition in the system so that we could immediately hear if the dolphins were reacting to or requesting toys in the wa-ter. So, software was designed and put into the hardware, and field test after field test in the Bahamas was made. This usually involved two researchers, who were wearing boxes that were hanging off a line behind the boat, playing sounds back and forth to each other for hours. First at ten feet, then twenty feet, then thirty feet, with the researchers at different angles to each other. We were also wearing bone-conducting headsets so we could hear what sounds were played from afar, and we could hear the label that the computer would give us. Adam would play "scarf," and hopefully I would hear "scarf" and report back to the data loggers onboard how correct the computer responses were. Sometimes Adam played "scarf," and the computer said it was "sargassum" (a type of seaweed). Wrong. So, boxes came out of the water, software got readjusted, and then back in the water we went. Testing went on for days, and anytime a major change to the system occurred (algorithms etc.), we retested in the water. I didn't want to expose the dolphins to the system unless it worked close to 100 percent of the time. But finally, it worked, and we were there.

One of the funniest stories that Thad tells is of an incident that occurred one testing day. Adam Pack and I were in the water, float-ing at the surface with our CHAT boxes on, playing sounds back and forth. We always tested in shallow water, and it's a relatively safe place to work because of the underwater visibility. We weren't quite sure if the electrical current from the boxes would attract sharks, so to be safe, we had our first mate, John Rose, as a look out in the water. Suddenly, John popped his head up and shouted to Thad on the boat, "That's the largest shark I have ever seen!" He then proceeded to dive back under the water, presumably to chase the shark away. At this point, Thad was on the boat thinking, "Oh my god, what if I just put in harm's way two of the world's leading marine mammal researchers." After hearing this at the surface of the water as we saw John dive under, we found out it was just a nurse shark—really not dangerous at all—so everything was fine, but it gave Thad a few heart palpitations. During our weekly Zoom meetings, I often hear discussion about "not blowing Denise up" from exploding lithium batteries in the box, so I guess safety concerns are both above and below the water.

With our newly tested CHAT systems, we used the model/rival technique, with multiple CHAT units in the water, to show the free-ranging dolphins the contingencies of this system. Four objects were labeled with synthesized whistles that the dolphins could mimic. Since we knew that dolphins greeted each other with their unique signature whistles, we gave ourselves signature whistles and put them in the system. We also put some of the dolphin's signature whistles in the system, focusing on the individuals we knew well and who work with us the most in the field. Now that we knew more about the etiquette of greeting a dolphin with a signature whistle, we began to greet specific dolphins when they approached. These additions allowed the system to function like dolphins do (at least to some degree—e.g., greeting each other with their signature whistles). Dolphins could hear the outputted sounds (as whistles) when humans used the system, and they could watch the exchange of toys between humans (the contingencies of the system), thus exposing the dolphins to a communication system. When one diver pushed a key on a keypad or if a dolphin mimicked a whistle accurately, the computer system recognized the whistle in real time and produced an English equivalent into the earphone worn by another diver. The diver heard what toy was requested and provided it or reiterated the label of the toy to the dolphin or the requesting human. Eventually incorporating high-frequency recording in the system (192 kHz), we were now close to "being" a dolphin, at least in the sound world. The system had all the advantages of a computer: it was fast, data was fully recorded, and programming changes could be made in the field if needed. Data could even be immediately uploaded via Wi-Fi when the CHAT units exited the water. Working in the field adds challenges you don't experience with tank-side interfaces. Weather, getting enough exposure with the same dolphins, and underwater computers in salt water all compound technical and user issues.

It is now 2013, and we are finally set to get CHAT in full motion for a whole field season. I get a call from Captain Scotty Smith on another dive vessel, who says, "The dolphins are gone, and we haven't seen any for weeks." Gone? Just gone? Scotty has been out almost as many decades as I have, and I know he has a sharp eye and is very knowledgeable about the area, so I'm worried. Sure

Figure 4.4. Celeste Mason, part of the Georgia Tech team, wears a CHAT box to demonstrate its elements. Photograph courtesy of the Wild Dolphin Project.

enough, we start our field season in May, and wow, we really can't find our main group, although we do see a couple other smaller groups around. There are at least fifty-two dolphins missing from our community. After a couple months of looking, we decide to head south to the next sandbank, called Great Bahama Bank, and a location called Bimini. Right before we leave, Scotty calls to say that he thinks he saw Little Gash, one of our dolphins, down in Bimini. Sure enough, our dolphins are there. And thanks to Scotty and Gene Flipse, another boat operator in the area, sharing their photographs, we can now account for all fifty-two of our dolphins.

This major displacement had left the dolphins a bit discombobulated, so starting our CHAT work in this situation seemed risky, since we might not have their full attention. After looking at the oceanographic features in the area, we determined that there had been a food crash, and the dolphins had to move to find food. In August we tried a bit of time with the dolphins and CHAT, but it was clear we would need to wait until next year's summer field season to really start the work.

The following year was a better year for CHAT, although we had decided to now cover both our old field site and our new field site during the summer. This meant less time at each site. But it worked for the most part. In 2015 I made a risky decision. Some of our best dolphin players who were interested in CHAT were in the most remote region of our original field site. One of the challenges with two-way work in the wild, at least the way we had designed our work, is getting enough exposure with the same individuals to give them a lot of time to see the system work. So, I decided to spend most of our time in 2015 way up north in the remote zone of Little Bahama Bank with our small, but curious, group of spotted dolphins. And it was a great decision. We had hours and hours with the same individuals, who often came by the boat while we were anchored to explore CHAT with us. When the dolphins came to our anchored vessel, it was a good indication that they were coming by to see us, as there is nothing much else around. Because the boat was anchored, we had less engine noise to deal with, and safety issues were easy to control as well.

After a full season of CHAT work, we had some intriguing results. As is often the case with science, sometimes what you are looking for is not what you find. We did find the dolphins tried multiple ways of mimicking our synthesized whistles. Sometimes they would tag their own whistle onto the artificial whistle. Sometimes they mimicked our whistles in a higher frequency. They mimicked numerically; they would produce three whistles after we produced three whistles. The most interesting result was found in the high-frequency data, where we observed that the dolphins were not

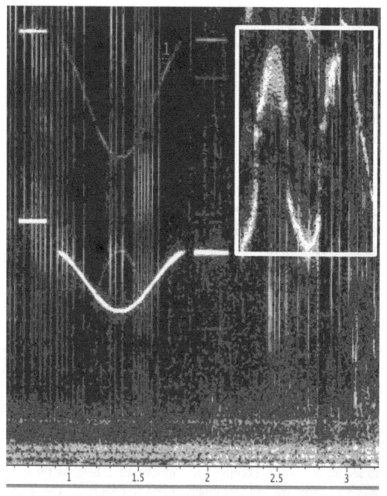

Figure 4.5. An example of a dolphin's response to a computer sound. The first half of the spectrogram represents the computer sound for "scarf." The boxed-in area represents the dolphin's response to the computer sound, in this case the dolphin's signature whistle. Courtesy of the Wild Dolphin Project.

mimicking our whistles in the way that had been described in captive work, that is by using tonal whistles. They were mimicking the shape of the whistle (the contour) through some complex manipulation of their clicks. And as my colleague Adam Pack noted, the dolphins just might have been showing us their preferred way of mimicking our whistles. Even in the best of situations in the wild, with very familiar dolphins, we never quite saw the dolphins grasp the functional meaning of the whistles in order to request different toys. Although surprising, it may simply mean that they have other interests in the wild more important than our interactions.

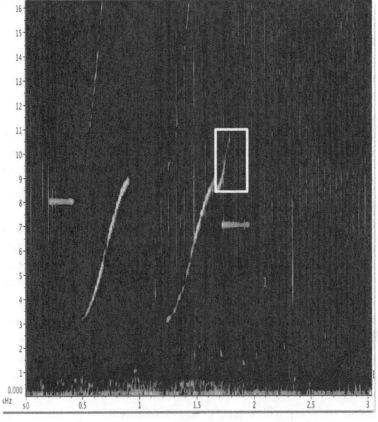

Figure 4.6. Dolphins mimicked parts of the computer sounds. The first half of the spectrogram represents the computer sound for "sargassum." The boxed-in area, an upswept whistle, was attached by a dolphin to the last upswept whistle that the computer generated, even before the complete computer signal was finished. Courtesy of the Wild Dolphin Project.

Or it could mean that they still didn't have enough exposure to our system. Maybe the toys were not motivating enough. Or perhaps we need to rethink our methods. Either way, we will continue to work on the system to give them more exposure and increase our understanding of their responses.

My hope is that if we can find enough interesting patterns in the dolphins' own communication signals, perhaps even language-like structures, we may be able to play the patterns back and see if this establishes another level of communication and to test the function of their sounds. For example, if we found a specific sequence of sounds when the dolphins dive together to forage, we could play this back and see if they engaged in this behavior.

Although my work has been focused on observing natural behavior and communication between dolphins, I have always wondered what the dolphins were thinking, and if it ever might be possible to communicate with them in a direct way. Historically, the only way to accomplish this has been to work with animals in captivity, and in an experimental way. Even the most productive cognitive experiments have always been highly controlled, one-way experiments, focusing on what the dolphins can understand not on creating a two-way communication system so they could talk back. As we have already noted, given the propensity of dolphins and other species to either use referential signals in the wild or artificial systems in the lab, pursuing language interfaces seemed reasonable. We also know that mimicking a signal is not the same as understanding its function. Vocal object labels can be imitated prior to an understanding of the referring function of these symbols. The referential function of such labels can be ultimately understood in different contexts, suggesting that allowing an animal the option to cognitively explore the functional use of such mimics is desirable.

Meanwhile, we were simultaneously working on what's called machine learning with our acoustic database. Eventually we started working with a great group of volunteers at Google to help us further our acoustic categorization and other aspects of pattern recognition of sounds. This may seem an easy task given today's machine-learning tools for human language—like ChatGPT—but I can tell you it is not as easy as it might sound. Even today, the

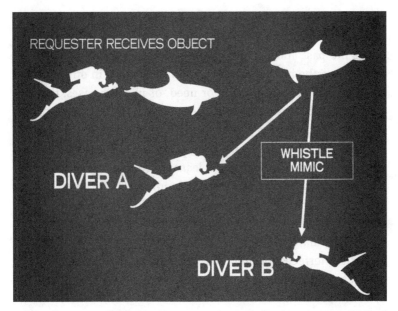

Figure 4.7. Divers demonstrate the CHAT system to an observing dolphin. Initially, Diver A plays a computer whistle. Diver B hears it as an English word; the dolphin hears it as a whistle. Diver B provides diver A with the requested toy. If the dolphin chooses to mimic the whistle, one of the divers rewards the dolphin with the toy, and they play and interact. Courtesy of the Wild Dolphin Project.

need for expert interjection (like my knowledge of dolphin sound types—called supervised learning) can really help the computer learn what to label. Throwing a bunch of sounds into a computer (called unsupervised learning) and hoping it will pull out biologically significant patterns does not work so well yet. Of course, Google, Facebook, and other companies have made great strides with this because of their massive access to things like facial data and human words. But to build these good models, you need a lot of data, and most biologists have nowhere near that amount of data to work with. But we carried on exploring the best way to find patterns in our data. Working hand in hand with Thad Starner's previous PhD student, Daniel Kohlsdorf, we have been developing a model for our dolphin sounds and applying and exploring sequences of sound to look for language-like structures.[31] With new tools and software, I suspect that, if applied to other species, we will begin to see that animals have at least the beginnings of some sort of structured language. After all, what could be more helpful in survival than being able to communicate in some detail, and about the past and present. I also hold that, besides the lack of technology, the human block to finding language in other species is simply a lack of motivation or need, or in some cases, the concern that such a discovery will knock humans off the pedestal.

I see two paths for overcoming those hurdles. The first is observing, recording, and analyzing communication signals in the wild. The new tools of machine learning will greatly enhance our abilities to make sense of complex communication signals of a variety of animals. The second way is to develop a mutual communication system between species, as we have reviewed already. And we already have physical keyboard interfaces for chimpanzees, acoustic interfaces for dolphins, and physical interfaces for dogs, who can pull and bite to communicate back to us. Both approaches can enhance each other.

When I led Caroh, a female spotted dolphin, over to our underwater keyboard in 1998 and watched her hang patiently in the water as I pointed to the keys and to various toys to show her the system, I wondered how this would develop over the years. Thanks to better technology, we had soon become "dolphins" with our underwater CHAT units. And as I write this, we are miniaturizing the CHAT

Figure 4.8. Dr. Herzing offers a scarf, a favorite toy of the dolphins, to interested dolphins. A few of the dolphins would often swim away with their new toy, and they would sometimes even return it later. Photographs courtesy of the Wild Dolphin Project.

system to an armband unit and an off-the-shelf smartphone for whistle recognition, since technology has now made this possible. The need to adapt to new technologies only underscores the many challenges when working with any species, developing interfaces and keyboards that are species-appropriate, and trying to make sense of parts of their world they are willing to share with us.

5

Yesterday's Tools

Science builds on itself, eventually outgrowing its own methods.

How do you decipher the communication of another species? We have seen how other species listen, mimic, and cocreate understanding. Is this communication akin to human language? Do other animals use words? Do their sounds have structure or grammar? Although the definition of language can vary, it seems that we have not had the tools and technology to study animal communication adequately enough to determine what information is encoded and shared in their signals. Even after decades of observing dolphins, I have found decoding and deciphering their signals to be slow, although machine-learning techniques are now helping us along the way.

Our treatment of nonhuman animals has, for decades, catered to our belief of our human exceptionalism. And although humans are unique in many ways, so are other species. As I've said before—this is a hurdle to understanding. However, now we have a chance, and the tools, to take a good look at animal communication and see what, if any, language-like structures other animals might have.

As discussed, we currently know that versatile forms of animal communication do exist and may simply elude us because of our limited sensory perceptions and abilities, and our inadequate recording instruments and analysis tools. We will explore the traditional ways in which we have we tried to understand animal communication. We will then explore some of the cutting-edge techniques that have proved successful in human speech recognition and in the mining of large datasets. We all take for granted that our phones will recognize our words and that Google will almost instantly provide answers to our questions. These tools simply did not exist thirty years ago, and they weren't applied to the science of animal behavior until very recently.

In some ways, acoustic language, because of its ability to transmit over long distances, is still considered the "last" test for intelligence. But we have been unable to decipher acoustic patterns to look for language in other species. We know that many animal species use tools, some use labels for things in their environment, many have large brains and great memories, and some can solve complicated problems. We have only begun to look at whether some animals have a few features, or the stepping stones, of language, as scientists like Duane Rumbaugh and Sue Savage-Rumbaugh have argued for chimpanzees.[1] As discussed, animal behaviorist Con Slobodchikoff states that the alarm signals of the prairie dog, and the way they use their signals, is a language.[2] As we have noted, there are many "features" of language—strict requirements that humans have placed on determining what is, and is not, a language. These include time displacement (talking about yesterday or the future), abstract idea transmission, combinatorial signals (recombining parts of sound, like phonemes, to create words), recursiveness (placing phrases within phrases, although some argue this a rule rarely followed in human language), and others. But let's look at how organisms encode information, and what tools humans have used to decipher this encoding in the past.

Tools of the Trade

All organisms must produce signals and encode and decode information. Sensory systems on Earth are varied but often include chemical senses (smell, taste), mechanical senses (touch), visual senses (facial expression, postures, electromagnetic), and acoustic senses (sound, vibration). Some others might include magnetic senses, gravity senses, and the like, although we know less about these sensory systems. The type of senses an animal uses will also determine how information is encoded, transmitted, and decoded within and between organisms. Plants, bacteria, and other smaller organisms show a variety of complex sensory abilities, and some are well studied. But here we will focus on social species, for the most part, when looking at how other species communicate with each other.

We have many good examples of how nonhuman species modulate and encode information in their communication signals. Predator-specific alarm calls have been noted for decades. Can we derive any universal rules that might help decode currently unknown information from other signals? Environmental drivers of habitat (e.g., dense mediums, clear channels) may have contributed to the evolution of different repertoires within and between species. For example, one theory holds that dense mediums should produce discrete signals easily discernible, whereas open habitats are good for graded repertoires and have a higher likelihood that information is encoded in a complex way. Another theory suggests that social complexity drives the need to encode complex information. Although there is good evidence that higher social complexity is related to higher vocal complexity, some researchers conclude that this issue is not so simple and that, looking at a variety of species in similar habitats, environmental or social pressures may not hold as rules for driving complexity.

The challenges of deciphering the communication systems of another species are multilayered, and of course limited by both our senses, technology, and our datasets. Here we will look at some of the challenges and techniques that scientists have used to explore animal communication. We have not taken an adequate look previously, and we are only now developing the tools that can supersede our human analytical abilities and speed up the process of discovery in order to take a hard look at the question, Do animals have language?

Flash forward to 2015, and I am categorizing burst-pulse sounds using a software called UHURA (Unsupervised Harvesting and Utilization of Recognizable Acoustics), which was developed by Daniel Kohlsdorf and my colleagues at the Georgia Institute of

Technology.[3] (For non–Star Trek fans, Uhura was the officer in charge of communications onboard the starship *Enterprise*.) It has been a long road, and we continue to adjust our tools to our needs, even in this new world. But let's look at different levels of analysis that we have pursued as a field, as it will explain why we are still wondering if animals have language.

One thing to try to appreciate right off the bat is that there are many levels of understanding that need to take place before we can use the word *language* to describe a communication system. And in science, you do your best with the tools you have at the time. However, it may require a retroactive assessment in future years to realize how things may not be as they seemed. I will give examples of these levels from my own field in studying dolphin communication to try to illustrate these points.

Although dolphins are social mammals and have developed societies in similar ways to humans as social beings, they live in an aquatic world and have a highly developed sense of sound. They can hear, and produce, sounds well above human hearing, and they have very directional sound projection and hearing. In the early years of dolphin research (1950–1980), with dolphins both in captivity and in the wild, scientists collected and recorded vocalizations (usually in human hearing range) by dropping a hydrophone from the side of a tank or the side of a boat. Sounds were collected, categorized, and analyzed. Theories were made based on sound structures collected and perhaps some surface observations of the dolphins themselves.

Decades later, scientists now collect and record vocalizations and use high-frequency (in the dolphins' ultrasonic range) hydrophone arrays to collect sounds above our hearing range and to localize sounds (e.g., to tell who is vocalizing). In some cases, we use underwater videos and localization equipment to correlate sound with visual behaviors or with individuals to extract more details of dolphin-to-dolphin communication. A recent attempt by marine mammalogist Laela Sayigh and colleagues to record tagged short-finned pilot whales with data devices that simultaneously collect sound as well as location and depth information has produced interesting results.[4] But because dolphin sounds can be very directional, some of the old data collected for decades may represent partial whistles because the dolphin, unknowing to the researcher, was turned aside or away from the hydrophone, so only a clipped and partial signal was collected. So, we need to clarify the real repertoire of the dolphin using new tools. And sound correlations can now be more clearly matched with detailed behavior, such as foraging or social behavior, and sometimes even matched with individuals. This, of course, helps with our interpretation of how dolphins use their sounds, socially and otherwise.

First, we need to know, and understand, the sensory systems of our species, and we need to have adequate instrumentation to record within a sensory range. Second, although human beings are quite good at pattern recognition and categorizing, animal signals present subtleties that perhaps only a computer can categorize, recognize, and cluster together. Third, we need faster ways to mine our data, since large datasets might exist only to be limited by technological processing. Fourth, cutting-edge computer programs and algorithms can present the human, usually an expert in the data, with potential structures and orders to test and correlate with behavioral states of the animal or other metadata. The human injection, called supervised learning, can be inputted at any point to help the computer refine its algorithms according to expert overviews of the data. Fifth, once patterns are discovered, they can be tested to see if they are simply correlated with ritualized behavior (courtship, foraging) or whether they are context sensitive, suggesting that animals have patterns and rules within their communication streams. Finally, discovering the meaning and interpreting the function of communication signals in a society is probably the most challenging step and the most vulnerable to incorrect interpretation. Here, we may need to take a lesson from anthropology, since exploring human cultures had similar challenges.

Early anthropologists also started by using observation to correlate sound and behavior to understand the structure of a human society and its communication. Eventually, or accidentally, the anthropologists interacted with the society to try to understand the nuances of both the communication and its meaning within a cultural context or during ritualized behavior. The advantage we have with humans, if we can share some understanding of a language, is that we can inquire about how an individual might perceive something, although this approach also has its flaws and challenges (deception, self-deception, etc.). In animal behavioral science, it is more typical to use experimentation in a laboratory setting, to control the variables, and then manipulate some of the information to address the function of sounds. So, determining the function and meaning of sounds is probably the most difficult of the challenges since it is often bound by the cultural context of a human society or animal society. Playback of sounds may offer some insight to the function, but such playbacks that might ensure the correct context or individuals present are also difficult to create.

What about universals in animal communication, be they human or nonhuman? Are there any overlapping features that might add insight and meaning to a stream of information? In animal communication (even human), prosodic elements or relative information can give meaning. For example, the increasing intensity, say loudness, of a signal can show growing aggression or agitation. You know from my voice when I am angry. I may draw out words or use lower frequency tones to express my anger. The addition of my hand-waving to these spoken words can help emphasize the meaning of words and my increasing annoyance. This adds a visual signal to an acoustic one, creating a multimodal stream of information for the receiver. Finally, the fine-grained modulation of any one of many parameters in a signal (amplitude, frequency, duration) can add meaning. Frequency is a signal rate, or number of signals per time. I can repeat a word to you to increase the frequency of the type of word, or in the case of pitch, I can speak in a high pitch (greater units of hertz per sound). Amplitude is the loudness of a signal. I can make my voice louder or softer. Even those who are not communicating acoustically can still amplify other types of signals to emphasize a point. When using American Sign Language to communicate, for example, a person can modulate the "loudness" of their gestural sign by keeping it close to the body (quiet) or moving it far away from the body (loud). So, prosodic features themselves might be universal, in the sense that they can modify different signals (acoustic, visual) in the same way.

From the work by biologist Eugene Morton, we know that both birds and mammals follow similar structural rules, like using harsh, low-frequency growls to communicate aggression and tonal, high-frequency peeps to express fear and appeasement.[5] Subsequent research by behaviorist Péter Pongrácz and colleagues has verified these rules for human perception of dog barks.[6] So, analyzing prosodic information in any structural sound streams we glean from animals may be quite important. And viewing this from an anthropological parallel, having metadata along with communication signals, such as the sex, age, and relationships of those communicating, can start to illuminate the cultural meaning of words. By using these combined methods and tools, we can begin picking apart the details and potential complexity of an animal communication system. And not only can emotive or motivational subtleties be encoded within modulated signals, but behavioral contexts may also be encoded. Such subtleties have been described for domestic chicks by biologist Mamiko Koshiba and colleagues and in the different types of "buzzes" for echolocating bats, signaling either drinking or feeding activity, by biologist Danilo Russo and colleagues.[7] In some cases, geographic territory can be recognized by subtle differences in the vocal "trills" of the bearded seal, as described by behaviorist Isabelle Charrier and colleagues.[8] Let's take a look at methods used to study animal communication, both old and new, and their pros and cons.

We'll start with one of the oldest techniques used to study animal communication: observing behavior. Ethologists like Niko Tinbergen and Konrad Lorenz started by preliminarily observing and developing an ethogram, which is basically an animal's repertoire of behavior.[9] In our dolphin ethogram, we have a list of body postures, bubble blows, tail movements, and so on.[10] After an ethogram is created, we then design a protocol for sampling behavior. When we analyze our underwater video, we score these various behaviors in the order we observe them. For example, we might see an adult dolphin chasing another dolphin upside down. Then we see the chased dolphin do a tail slap to the head of the chasing dolphin. Basically, all these behaviors are coded over time, and we can start asking questions like, How many times did a dolphin tail slap another dolphin during a chase? Or we might ask, Do young dolphins tail slap more than older dolphins? It's really a way of trying to quantify dolphin behaviors. The most important thing about an ethogram is that it has an operational definition. For example, mating behavior might be defined as two individuals engaged in copulation. Or a head-to-head posture might be defined as two or more dolphins facing off head to head. Such operational definitions allow me to compare my study with other researchers so we can compare similarities or differences in communities or between species of dolphins.

The human eye of course has its own limitations, so for many species, it can be helpful to slow down or speed up video or sound recordings to score behavior. In my own work studying underwater behavior of fast-moving dolphins, it is usually imperative that we slow down video for analysis. It's amazing how many subtle visual signals can be seen on a slow-motion video. And without slow motion, we simply would not see the fourteen thousand wing beats per second of a hummingbird. Likewise, we can speed up a video to observe a plant growing and changing. So, time scales matter as well as sensory systems. Similarly, sound recordings can be manipulated in the analysis phase by slowing down a playback of often rapid sound information for the observer to score or categorize. And remember animals might communicate in frequencies beyond our perception, so unless we know mice produce ultrasonic vocalizations, we won't be listening for them or recording them. If an animal is making sounds above, or below, our hearing, spectrograms can identify these sounds that humans can't hear.

Developing an ethogram is a great place to start, giving you experience observing an animal, which then allows you to begin testing a theory or idea. How we create an ethogram can vary and have a great effect on the outcome. For example, you might choose to categorize only the very overt physical postures a species makes and ignore subtle body orientations. Or the opposite—you may choose to look at subtle details of physical contact while ignoring configuration challenges of groups of animals. These decisions are not easy but are usually guided by the questions a researcher wants to ask.

Another older technique is listening to animal sounds, as many animals use acoustic communication. As we have noted, it is one of the senses most able to transmit over long distance, both in the air and in the water. So recording, categorizing, and analyzing this aspect of communication behavior can be productive. But understanding acoustic communication can be complicated due to different ranges of acoustic sensory systems, the physics of sound traveling through different mediums, and the ability of humans, and computers, to even display or categorize sounds.

Again, we need the right tools to analyze sounds, since the human ear has its own hearing range and ways to lock into salient sound types. Typically, we use a microphone or an underwater hydrophone to record sounds. Equipment can be specialized toward directional collection, high-frequency collection, and other aspects to maximize our recording abilities across species. Of course, as technology has evolved over time, so has the scientist's ability to record sounds. Categorizing sounds can be done visually or by computer techniques that are unbiased by a human observer. Of course, understanding the variance in a signal is important. For example, I can say my name, Denise, in many ways. I can say it slowly or rapidly. I can say it with an inflection suggesting a question, or I can say it emphatically with a strong and quick emphasis. Much of this variance is conveying my mood or emotion; however, the word remains the same. So how are we to know when there is variance in an animal's sound whether the sound is the same or different? The grunts and groans of animals are easy to dismiss as just emotional or motivational states. But diligent scientific work (as with vervet monkey and prairie dog alarm calls, as noted previously) has shown us that when we are open to looking for referential signals or labels in the animals' world, we sometimes find them.

I have often floated in the water while taking a video of a group of dolphins fighting and wondered why sometimes the dolphins were very vocal, with a cacophony of sounds, and other times they were quiet while doing the same activity. But were they really quiet, or were they sometimes making high-frequency vocalizations that I just could not hear? Are they hiding from a predator, or another dolphin group, by being "silent"? As we have already noted, it is important to know a species' sensory systems. So, with dolphins, one of our main challenges is to record the ultrasound—vocalizations produced well above our hearing range (as with the sounds of bats and some rodents).

Early on, our only option was to have a specialized reel-to-reel tape recorder on the deck of the boat and to drop a hydrophone over the side. This would get us ultrasonic recordings but no underwater behavior to go with it. In my team's work, it has been critical to be mobile in the water with our equipment, since the dolphins rarely stay in one place for very long. Over the years, we began to use high-frequency recording equipment in the water, usually custom designed for our field situation.

In 2003 we started working with marine mammal acoustician Marc Lammers, then a student under the guidance of Whitlow Au, a well know dolphin sonar expert. Marc developed UDDAS (Underwater Dolphin Data Acquisition System), a mobile recording system for us to use in the water. The device housed a computer in the bottom of the unit, synchronized with an underwater video housing sitting on top. We were now able to swim

Figure 5.1. UDDAS (Underwater Dolphin Data Acquisition System) shown with the video camera housing on top of the main computer housing. Since the units were connected, this allowed us to synchronize the recordings of sounds and behavior. Photograph courtesy of the Wild Dolphin Project.

freely through the water and record ultrasonic sounds with our underwater video. This was new technology that revolutionized our work, giving us the ability to see and record in the ultrasonic range.

If you look at a typical spectrogram of sound from a UDDAS unit (fig. 5.2), you can see how the narrowband recordings compare to the broadband recording. Although humans can hear the narrowband (up to about 20 kHz), anything above that is considered ultrasonic. The whistles on the far right of figure 5.2 have some harmonics that can go into the ultrasonic area. Although we always knew that dolphins made clicks above our hearing range, in some cases there was no evidence of these clicks if you were

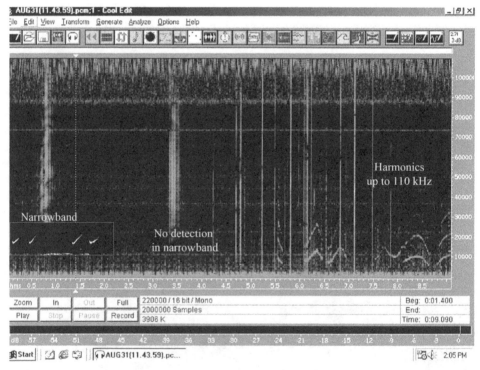

Figure 5.2a and 5.2b (inset). Dolphins produce sounds over ten times as high as humans. This spectrogram shows harmonics and clicks up to 110 kHz, well above human hearing range, which goes up to about 20 kHz. A comparison can be made with the spectrogram insert on the bottom left, showing our human hearing range. Courtesy of the Wild Dolphin Project.

only recording in a narrowband range. UDDAS finally allowed us to record both audible and inaudible sounds in the water while recording video.

In addition to recording the ultrasound, UDDAS allowed us to see how directional the dolphin signals really were. So, if we wanted to collect the full sound, we needed to be in front of the dolphin. Dolphins have both directional hearing and directional projection when making a sound, especially high-frequency sounds. If you collect sounds from behind or to the side of a dolphin, your recordings will quickly drop off both in intensity and in frequency. We could see this in our recordings; if a dolphin came head-on to UDDAS and then turned away, you would see the top of the whistle clipped on the recording as the dolphin swam away. Marc and others have proposed the theory that directional whistles, and their harmonics, function somewhat like an acoustic beam (like a flashlight beam, only sound). If a group of dolphins is swimming along and the leader turns, her echolocation clicks would suddenly be heard only in the lower frequencies, thus allowing the following dolphins to adjust their direction to follow the leader.

It turns out that much of our analysis equipment is designed for human systems. Marc pointed out that if you adjust a dolphin whistle for how a dolphin hears, instead of a human, it shows that the dolphin will hear the first or second harmonic better than the fundamental frequency (dolphins' best hearing is between about 30–40 kHz and 120–130 kHz—both above our hearing range). So, as humans, we need to be constantly aware that biases with our equipment and tools will also need to be adjusted to understand the dolphin world.

The last, and perhaps largest, challenge is identifying which dolphin is making a sound during a behavioral event. Since there are no reliable visual cues (other than occasional bubble streams synchronized with whistles), this problem needs to be solved to truly record and understand dolphin-to-dolphin conversations. It's the difference between recording humans talking in a conference room without visual information showing whose lips are moving and recording a video that clearly tracks who is talking. The challenge of localizing vocal communication may also apply to other sensory systems (e.g., chemical) in the sense of determining who is creating a signal. If I am studying visual signals, I can see who is making a signal, but sound presents a completely different issue. At a party I might be able to identify from the kitchen, which of my friends is talking in the other room, but that is only because I am familiar with their voices and expressions or laughter.

In the Bahamas we often record spectacular underwater video with large groups of dolphins fighting. But there are so many dolphins and so much going on, it's impossible to tell who is making the sounds. Although I had watched mothers and calves playing, and I had occasionally correlated a bubble stream produced by the mother calling her calf, these large groups of dolphins presented a much greater challenge. If we were to truly understand dolphin-to-dolphin communication, we needed some way of tracking who was making what sounds. On land, you can set up three microphones to localize a calling individual. But dolphins in the wild are mobile and do not stay centered in some floating hydrophone array. If only we had this technology for underwater.

After a few prototype experiments in localizing equipment, we finished testing a prototype in the Bahamas of an underwater localization system called ASPOD (Acoustic Source Positioning Overlay Device), which was custom designed and built by our colleague acoustician Matthias Hoffmann-Kuhnt, from the National University of Singapore.[11] His localization unit consists of a video camera with three hydrophones that collect sounds together.

We then post-process the acoustic and video data together, giving us a new rendered video showing which dolphins are vocalizing. The program puts a red square on a dolphin that is echolocating, a yellow star on a whistling dolphin, and so forth. We can now record dolphin-to-dolphin conversations and track the speaker. Of course, nothing is ever that easy. It turns out that because echolocation clicks are very short in duration, the computer can quickly locate them. On the other hand, whistles that can be a second long are harder to locate since the dolphin who is whistling continues swimming. The computer grabs the beginning of the whistle to locate the vocalizer, but by that time, the dolphin has moved through the water away from where it started the whistles. So, that is another problem to at least be aware of when reviewing the post-process videos. It may not be a problem when there are only two dolphins

Figure 5.3. Dr. Hoffmann-Kuhnt focuses ASPOD (Acoustic Source Positioning Overlay Device) on a curious dolphin to record and localize sound. Photograph courtesy of the Wild Dolphin Project.

on the video, say a mother and a calf reuniting. But with a large group of dolphins fighting, the system will need better resolution and tracking to discern the vocalizer. Although it will take years to collect enough data in different behavioral contexts to begin to understand some of the details of dolphin conversation, it's a start.

Audience effect is another important thing to consider when capturing behavior. Do the producers of signals adjust their signals depending on who is around? If friends are around, do they share their excitement about finding a good source of food? Or if foes are around, do they try to deceive them and hide the location of a new-found food source? Audience effect is quite prevalent in the animal world, and it demonstrates how cognizant animals are of their environment, their relationships, and the implications of their behavior and actions in a situation. Researcher John Mahoney and colleagues describe crypticity (the hiding of a signal) as a complicated process.[12] The demonstration of covert communications is

08-18-2019 Sun 11:24:06

Figure 5.4. The finished, post-produced video labels echolocating dolphins with a square and whistling dolphins with a star. This allows researchers to analyze how dolphins converse with each other. Screen results courtesy of Matthias Hoffmann-Kuhnt.

a striking example of animals understanding who a potential receiver of information is and adjusting their communication signals accordingly. The evolution of hiding communication signals is discussed extensively by Gareth Jones.[13]

If you want to eventually interpret communication signals, it is important to have other information about the individuals or society you are studying. In my own dolphin research, my team and I have a long-term database of who's who—who is related to whom, who sires calves, and who are friends. These important life-history details, along with age and sex, are called metadata. This refers to the extra information that a researcher might collect to further understand behavior. We collect sounds, videos of visual behavior, individual identity, sex, relationships of the group, and so forth. These layers of history are critical when I want to look at the meaning of a communication signal. As described before, I might see one dolphin who is upside down chasing another dolphin. If I know that the inverted dolphin is a mother chasing her calf, I can interpret her action as a type of discipline and follow the chase through to see how the mother enforces rules designed for safety and survival. If, instead, I see that the inverted dolphin is a male of reproductive age, and he is chasing a known female dolphin also of sexual maturity, I can see that this is a mating attempt, and I can follow the chase to see whether the male is successful in mating, or alternatively, I can see the female dolphin slap the male with her tail, in an apparent signal of noninterest. In both cases, mother/calf and male/female, the physical posture is the same, but the players are different. Following the process of a stream of information can inform and validate the context and function of signals (discipline versus mating).

With metadata, these events can be separated by their meaning and interpreted differently. Perhaps what we call metadata is just part of the larger encoding or signal to the species, but as human observers, we often separate these events into mechanical moves (physical postures, etc.) and players. There are many types of metadata, and probably the most difficult to gather is historical information. How often did one animal get chased by a predator? Did it witness its sibling being killed? Did it have trouble finding food the last drought? All these historical factors can play into an animal's behavior and are probably the reason there are variations in behavioral patterns that challenge our interpretations. Another aspect of metadata that may add to the variation is personality, now well studied for many taxa. Although the personalities and relationships of animals in their own groups can make studies more difficult, it is exactly this element that we are trying to understand. As anthropologists H. Russell Bernard and Clarence Gravlee suggest, metadata is perhaps an anthropologist's (and a behavioral biologist's) most powerful tool.[14]

Even if we discover complex order and structure in an animal communication system, we still need to try to interpret the meaning of the signals. We want to answer one big question: Can animals communicate about things from the past and into the future? We know that many animals can store food or recall a watering hole over years, but can they talk about these things with their comrades, or do they just follow an action, like digging up their acorns or leading their pack to the watering hole without discussion? If animals plan, or have a sense of the future, would there be a need to communicate such things to others? Would there be an evolutionary advantage for survival to be able to communicate such knowledge? Undoubtedly many of the more complicated actions would require language.

Multimodal Communication

Understanding the total sensory information stream during a communication sequence is critical. Sensory cues can work together to enhance signals. Humans don't limit themselves to one sensory modality when communicating to another person, unless they are restricted during long-distance communication (e.g., via phone or email). If they are, subtle body signals will be missed. The use of video calls now allows us to see our long-distance friend and gain more information. The same is likely true during communication in other species. I am a big believer of observing in the wild and manually processing or reviewing sounds, even when using a computerized system. There is nothing that replaces exposure to the real thing and knowing your data. Computers can spit out incorrect results for many reasons, including data-entry errors or programming errors. So, the ability to know your data somewhat intuitively can help catch these technology errors. And perhaps understanding one modality well can help with the process of overlaying multiple modalities.

Although it can be difficult and more time consuming, we must be cognizant to not limit our processing of communication signals to just one modality, or to our human perceptual abilities. In my own work, I have always focused on correlating underwater visual behavior with sound. And as soon as the equipment became field-feasible, I added high-frequency recording systems to capture the high-frequency nature of dolphin sounds. Although I analyze my sounds separately, I rely on visual data, and the metadata, to interpret a sequence of acoustic behavioral activity.

The way human researchers try to understand most animal behavior is by correlating sound and visual observations. Human behavior, with its complexities, is also described using more than one modality. As discussed by Donald Favareau, the excellent work by behaviorist Chuck Goodwin and others brought this realization into the analysis of human communication.[15] So, too, can ethologists have both visual and acoustic data in their tool bags to help interpret animal behavior. But there are some serious obstacles that make it extra challenging, and we need to consider these in the process.

What if animals use a couple different modalities to express themselves, and what if we are recording only one of the modalities? This is analogous to you recording my voice telling a story, instead of taking a video of me telling a story. You might miss my hand waves, my body postures, and my eyebrow nuances. These are all details of signals that can and do matter in interpretation of behavior. The fact that these elements are missing from email and text communication has presented a challenge, and so the emoji was born to help us communicate our feelings or emphasize a state of mind.

To study an animal, you might first observe its behavior and gather a full repertoire of its signals. For example, I might first list the various types of body postures, using an ethogram (labeled movements, like chase or pec to pec), that occur in a short video clip. But because dolphins are very acoustic, I also label sound types that are present (whistles, clicks, etc.) during these physical behaviors. This allows me to see the interplay between body movements and vocalizations. Although some of the sounds are overt and easily labeled, some are not. I might then also note categories of sounds that have been identified by my machine-learning program, since I can't categorize all of them visually without the help of the computer. So now I have all my sound types matching what is going on with the video. And if my AI programs are working well (we will look at this in the chapter ahead), I might even see some repetitive sounds and structures of sounds, which suggests word order and rules in their vocalizations. If I have my localization equipment in the water at the same time, I should be able to label *who* exactly is making what sound during these physical behaviors, and this gives me more detailed information.

Now, say I have the same stream of body postures and sounds for multiple video clips, except for a few different sound types. I am wondering why, in some of the videos, there are a few different sounds within the larger pattern of sounds. If I look at my metadata, to see *who* was there, *what* their relationships were, and *what* their age and sex is, I see that in the one example described, I have an adult male chasing an adult female, in the behavioral context of courtship/mating. In the other example, I have a mother chasing her offspring in the context of discipline. The sounds that are different in the second example consist of a couple different types of whistles (which I have previously determined to be signature whistles of different mothers and their calves). So perhaps the sound sequence is the same, except for the whistle substitutions (we will look at this issue in the next chapter, when we explore the use of alignment-learning algorithms). In this way, I can put all the information together to help me interpret signals.

To fully understand what dolphins are saying to each other, we use a tiered approached. Much like with an anthropological study, by knowing who the dolphins are, their sex, and how they are related, we can more accurately interpret their behaviors and sounds. By knowing the individuals and their relationships, we can describe the example given when a mother disciplines a calf. In the other video, where we also know the individuals and their relationships and sex, we see the male upside down chasing a female to mate, and our interpretation would be quite different.

Scientists often use playback as a way of testing the perception of information or the function of a signal. For example, as we've discussed, in the late 1980s biologist Robert Seyfarth and colleagues used playback to confirm that vervet monkeys (previously recorded in predation situations) responded appropriately to alarm calls for leopards, eagles, and snakes.[16] Each successful escape from a predator required a different reaction, like running up a tree or hiding under a bush. If a general sound is made as an alarm—for example, me yelling "look out" or simply screaming in a room full of people—no one would know where to look for the threat or what action to take. Thus, labels become critical to survival, and as Con Slobodchikoff and colleagues have pointed out, alarm calls are a first place to look for "words," since a lot of specific information can be compacted into a short amount of time, out of necessity to increase survival.[17]

Playing back recorded sounds of animals to test their reactions gives us solid evidence of the potential "function" of these sounds. A well-designed experiment can do what we cannot do by simple

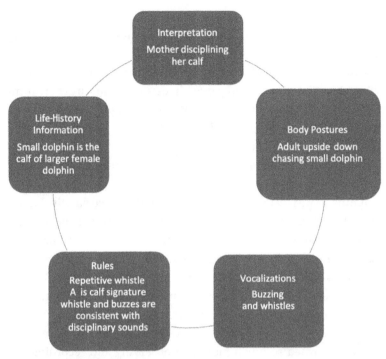

Figure 5.5. There are layers of information that help us interpret what is happening in our underwater video and in our sound analysis. Courtesy of the Wild Dolphin Project.

correlation—it can test a specific function (although this can be tricky if you can't see the animal's reaction, like underwater, for example). This then can feed back into the deciphering of natural sounds, since it gives us a subset of "calls," for example, that we may then examine for modulated features like frequency, duration, and amplitude.

Historically this technique has worked for terrestrial species, who must be painstakingly studied to generate the specific questions and acoustic signals to play back. Slobodchikoff's work, like Seyfarth and Cheney's, used extensive acoustic analyses, as well as experimental designs, to test the prairie dogs' reactions to a variety of stimuli. These research projects took decades, thirty years or more, to discover the use of alarm calls, and much of the acoustic work might now be accomplished by modern day tools, like the computer and pattern recognition programs. Unfortunately, such tests are difficult to do with an aquatic species. Testing the "behavioral response" of whales to US Navy sonar, for example, is nearly impossible, since you can't really see what the whales are doing under the water, although researchers have tried and at least noted things like directional changes. Such tests may show us short-term responses, like they avoid the speaker or change their distance as they migrate past the playback area, but we are unable to observe the whales' underwater behavior in any detail.

As described previously, earlier in my studies I had tried playing back some signature whistles I had cataloged from the spotted dolphins. It was clear from that experiment that we simply did not know the etiquette around signature whistles. Was it inappropriate to play someone's signature whistle when they were present? Signature whistles are basically names of individuals. Animal behaviorists Nicola Quick and Vincent Janik have shown that dolphin groups meeting in the wild may produce a group's whistle but add a bit of their own whistle on the end, perhaps identifying the caller.[18] After my initial experiment, I decided not to do playback experiments. I learned so much more just observing and recording the dolphin's behavior. Now, with our new CHAT system, I felt we were ready to selectively use signature whistles to greet the dolphins we were working with. And there may come a time, once we know more about specific words or sequences of their communication, when it will be appropriate to try playback experiments with dolphins again. Luckily, we have many new techniques emerging in the computer sciences that can help us with these complicated problems.

6

Tomorrow's Rosetta Stone

"Google translate" for deciphering other species—is it coming?

In 1988 I was working on my master's degree, and I was getting frustrated. I had taken on the challenge of measuring and trying to categorize burst-pulse sounds. These are sounds made up of very densely packed click trains (imagine running your finger quicky across the teeth of a thin comb; that is like a burst-pulse sound). Dolphins use these sounds more than whistles, and primarily when they are in proximity to each other and fighting. Unlike whistles, which have been studied and categorized since the 1960s, burst-pulse sounds are not well studied. This is because they are hard to categorize on a spectrogram using the human eye. Researchers had always focused on whistles because they are easy to recognize, and it is easy to measure their frequency changes over time. But burst-pulse sounds are messy and tend to merge into each other, making them hard to categorize.

Burst pulses grade into each other, change click rates and amplitudes, and need every small measurement to categorize. Although their temporal characteristics are quite measurable (interval, the space between signals, etc.), everything else, like how loud certain areas of the frequency spectrum are, is variable. I wondered how the dolphins categorize them and if there would ever be a software program to help us categorize these challenging dolphins sounds.

In recent decades, we have seen a rapid transformation in information processing. Although some software methods and tools have been around for a couple decades, the rapid application of these tools to large datasets (e.g., Facebook visual data, Google voice recognition) have improved old methods and created new tools for large data mining. But, as discussed in the previous chapter, animal behaviorists still use many traditional means for decoding signals. These include basic programs that measure things like frequency and duration (they go by names like "discriminant function analysis" and "principal component analysis"). Recently developed tools fall under the umbrella of artificial intelligence, including neural networks, machine learning, discovery algorithms, and deep learning. But before AI, researchers still measured the basic structure of sounds to make sense of them. If you are not interested in some of the more technical details of current AI, you might skip the rest of this chapter. But I have tried to make it somewhat digestible, using examples from my own dolphin work, in case you are curious.

And although human judges can give us a qualitatively relative comparison of whistles, it does not give us a quantitative measure of how different or similar whistles are to each other. Historically, in the field of matching signature whistles with individual dolphins, researchers have mostly relied on visual inspection of an image, enlisting human judges to match whistle contours. It turns out humans are good visual pattern recognizers, as we can see with the four distinct signature whistles in figure 6.1.

So now I knew that some whistles look pretty different, some are more similar than others, and so on. But I wondered how similar or different signature whistles were between mothers and calves. They kind of look like they have the same shape, but it is hard to measure. Do calves make longer whistles, for example, or do they even look alike at all in related individuals? By using a neural network, we can train a computer to recognize each dolphin's signature whistle and generate an actual measure, or index, of *how* different the whistles are.

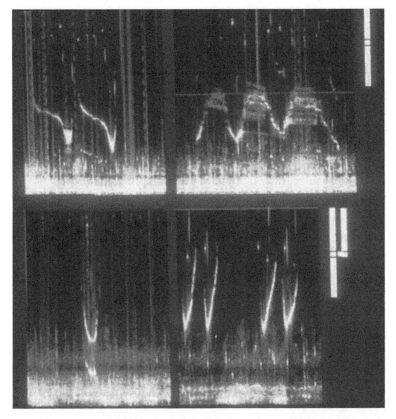

Figure 6.1. Four distinct signature whistles from four different dolphins are shown. Humans are good at recognizing visual patterns, and we can clearly distinguish the differences between the four whistles. Courtesy of the Wild Dolphin Project.

In my early years, I worked with Volker Deecke, an acoustic scientist who began writing programs to use neural networks to look at dolphin whistles.[1] We grabbed twenty-five examples of one of our dolphins' signature whistles and fed them into the computer to train the computer. Then we grabbed another twenty-five whistles from another dolphin and fed them into the computer. After more individuals were entered, we then let the computer see how well it could match all the whistles. In the case of a mother and calf, the computer showed completely overlapping signatures whistles, thus the computer program could not discriminate well between those whistles. In the case of two unrelated juveniles, the whistles were quite different and easily separated by the computer. So not only had the neural network showed us what was similar and dissimilar (which the human eye can do pretty well), it also generated a numerical index, or measure, of *how* different the whistles were, giving us a quantitative measure.

One thing that became quite clear in working with neural nets was their limitations. Some of our dolphins had what I would describe as "noisy" whistles. These were whistles that had a burst-pulse component simultaneously produced, and since the neural network required us to "trace the contour" of the whistle, messy whistles just didn't work. So we couldn't use some of the signature whistles in our data because of their messiness.

What was even more interesting was that as I began mentioning this to colleagues, I found that many of them also had "messy" whistles in their datasets and just threw them out (well, put them in an unused folder, since scientists avoid throwing out any data). And as the scientists relooked at these messy whistles, they realized they were not noise at all but significant sounds. So, these types of whistles really became a sound type, not just a messy whistle, and they ended up being valuable for future work. These "messy" sounds are a good reminder to ask ourselves, What is a signal, and what is noise? And this is also an example of how having an analysis program that restricts your data entry could in turn restrict or bias your results to "clean" whistles only.

Scientists quantify signature whistles in basic ways, so like others, I manually measured the whistles and generated data about them. Some standard things to measure in a whistle include the length of the whistle, the maximum frequency or pitch, and the minimum frequency. I can then enter my measured variables into a statistical program, in this case one that uses discriminant function analysis (DFA). DFA is a statistical tool used to predict categories of variables by grouping data by dependent variables. It helps us determine whether a set of variables can predict membership into a class. It's usually used when groups are known ahead of time, and it basically helps us put things into classes or categories of similar type. By using DFA, I get whistles that cluster into relatively separated clumps by duration and frequency. Or I can use principal component analysis (PCA), which also reveals clumps of similar types of whistles and indicates which dimensions are most important. PCA, also called Karhunen-Loeve transform (KLT) in signal processing, tells me which variables are most important for clustering whistle types.

A hidden Markov model is another technique that can tell you what the most important measurements of your sound are. This method has been around for a while. If I put some dolphin whistles in my hidden Markov model, it might reveal that duration is most important to the character of these whistles rather than frequency. There are many mathematical models that can help sort out sound types and the important features that they contain.

Over the years, after measuring and comparing all my signature whistles, I had a pretty good idea what a calf named Nassau's whistle looked like, as compared to the whistle of her mom, Nippy, and others. But I started noticing that sometimes Nassau's whistle would have a longer duration when she was excited. Or more loops when she swam rapidly. I noticed that Luna's whistle would sometimes be shorter when she was chasing her newborn calf, Lanai. Would the computer recognize these whistles as being the same signature whistle? What if the whistles were stretched out over time? What if dolphins changed their whistles when they were excited or scared? We know humans can compress or expand a word in time, giving it potentially different meanings but still representing the same word. A handy technique for handling this problem is called dynamic time warping (DTW). DTW is an algorithm that measures similarity between two signals that may vary in duration. This method essentially takes a couple whistles and matches peak points within the whistle, even if they are stretched out or compressed. So, if a mother is stretching out a whistle more during one behavior, it can be matched to her normal whistle or when her whistle is shorter and compressed. The same thing happens if I call out "Zorro!" to my cat rapidly if he is misbehaving versus "Zorroooo" if I want to give him a treat. The word is the same, but I have either compressed the word or stretched it out. Either way, it is the same word. Similarly, DTW can compare dolphin whistles that might be the same but just compressed or stretched out in time.

Remember that it is possible that this stretching or compressing is also expressing important information beyond the whistle itself.

For dolphin research and other animal communication research, this means we have a technique that will let us compare the same acoustic signal produced in different behavioral contexts. We can ask questions like, Do mothers produce longer signature whistles when their calf is out of sight? Or, Do dolphins make shorter whistles when they are being chased by a predator? This is the acoustic equivalent of an animal speeding up or slowing down their body movements to communicate urgency or a relaxed state. So similar things can be communicated in different senses by modulating the speed of a signal. And in certain human cultures, hand-waving and rapid words both communicate similar things.

All these techniques gave us powerful tools to sort out and measure dolphin sounds. But now I want to start looking at sequences of the dolphins' sounds and whether they have any order or structure to them, with the idea of looking for language-like structures in their sound streams. Every language has some kind of order and structure. And every language has complex units of information that are balanced with an efficient number of units that can be recombined. So how do we start looking for such patterns? In the past, a well-known approach was to look at a Markov chain of signals. This technique lets me look at how many times whistle A is followed by whistle B (this would be the equivalent of looking at how often one English word follows another). Or how many times behavior A is followed by behavior B, and so forth. But it doesn't get very far if you want to look at complex chains or strings of signals, and I wanted to look at complex sequences.

The Power of AI Tools

Enter artificial intelligence, which we consider as part of our new world. And there are some awesome techniques that have propelled us ahead with human language and cross-translation. These are the things we currently take for granted during our Google searches, Facebook entries, and use of a myriad of other computer features, including digital assistants like Siri and Alexa. How did Google do it? How did Amazon do it? Well, one of the first things to remember in any machine-learning process is that you need a lot of data, and I mean millions of images or sound bites.

I had already started mining and using machine learning with our dolphin data since Daniel Kohlsdorf had developed techniques to help cluster our dolphin vocalization data. We could put in a sound file (or multiple sound files), and after some tweaking, the computer would generate clusters of sounds that were similar. This is essentially done through machine learning. So now I had thrown

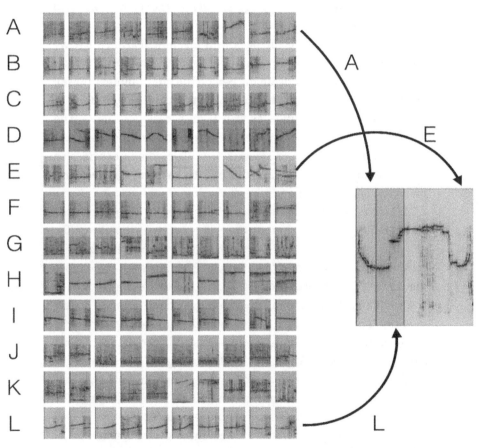

Figure 6.2. Cluster analysis using machine learning can help us find similar sounds. Each row represents a cluster of sounds that look similar to the computer, as generated from our UHURA (Unsupervised Harvesting and Utilization of Recognizable Acoustics) machine-learning program. Once categories are labeled, we can look at how they might recombine to form different whistles, like "ALE" in this example. Courtesy of Daniel Kohlsdorf.

a bunch of dolphin sound files into a machine-learning program, with a bit of human feedback, and the computer had come up with categories using a combination of the above techniques. This is called cluster analysis.

Cluster analysis is a tool used when we don't know what groups exist, and it is often used in exploratory data mining. Cluster analysis can use various algorithms to achieve its task, or a combination of a neural network and a hidden Markov model, and the design of such algorithms can evolve depending on what the scientist thinks may constitute a cluster, or category, and how the scientist wants to find or optimize a search for clusters.

But there are some challenges with machine learning. Say we are recording a human conversation. I set down my coffee cup on a glass table (noise) and drum my pen on the table (another noise), all while speaking words. The computer may cluster my coffee cup sounds and my pen sounds, as well as an occasional sneeze. The computer doesn't know what the functions of these sounds are, but it groups them separately, nevertheless. The computer will also cluster all my words, including those with sharp onsets, lower frequencies, and so on, or any other individual voices.

If I have metadata, say a video of the meeting where I can see who is talking, I can go through the clusters and see that the coffee cup and pen drumming noises are just other sounds that are not significant in the stream of "real words." But they were sound types that the computer recognized and categorized. We might go back later and cluster my coffee cup sound and the drumming collectively as "noise"; a listener might argue that my pen drumming is significant, in that it is observed while a specific person, who I don't agree with, is talking. Thus, my pen drumming becomes a modifier for the whole conversation and potentially important information. In some ways, my pen drumming might be considered a prosodic element of emphasis, holding emotional significance.

Back to the dolphin data. Since we want to know about the structure of an entire sequence (not just "sound A is more likely to follow sound B," which is an old Markov chain technique but very limiting for large sequential analysis), we want to know if there are patterns within patterns, language-like structures, or evidence of recombining small units of sounds that might hint at potential language. This is where it gets both difficult and exciting. These new techniques are evolving every day. And thanks to large datasets of visual information (Facebook) and human voice information (Google), we have some cool new techniques to apply to animal communication data.

In my human example, I decide to tell the computer to ignore all my coffee cup sounds and my pen drumming sounds (as noise), and I cluster all the "real words." The computer can then go through and tell me about the order of the words, or the rules (grammar) of my sentences. One of the first things about language you find is that there is a semblance of order. Although word order may be different between human languages (e.g., "This is Mary's house" in English versus *"Esta es la casa de María,"* or "This is the house of Mary," in Spanish), it still takes order and patterns to create a grammar and ultimately a language. So now the computer has figured out what clusters of sounds I am interested in (after my supervised input), which sounds I believe to be insignificant noise, and how the real sounds are clustered and used structurally. In the case of dolphin data, I can tell the computer to ignore all the clusters that are "noise" (boat noise, snorkeling noise, etc.) and just show me the clusters that are dolphin sounds. This is where the expert, or supervised, learning comes in. The computer will search for whatever sound types are there, and if left unattended (unsupervised learning), the computer may label a cluster of noise along with other relevant dolphin sounds. So semisupervised learning may be helpful to guide the computer at a certain level.

One thing we have discovered with unsupervised learning, in this example, is that the computer may pick up on something in the background and decide to match based on this feature. But what really needs to be matched is in the foreground. So, you never know what the computer might be matching until you look yourself and verify it. Computer scientist Scott Sorensen and colleagues describe an example where, instead of matching individual polar bears in the snow, the computer started clustering types of snow.[2]

So, after "supervising" the computer for a bit to make sure it is giving us the information we are interested in (dolphin sounds not boat noise, for example), we can start looking at sequences of sound types. Now if I want to know whether dolphins recombine various parts of a whistle to make up different whistles, I can look at that.

Figure 6.3. When polar bear photographs were input into an early machine-learning program, the computer picked up the types of snow in the environment not the individual identity of the bears. Supervised learning can help guide the computer as to what is salient for the researcher to discover. Photograph courtesy of Andreas Weith / Creative Commons Attribution-ShareAlike 4.0 license.

Historically, scientists have measured a dolphin whistle as a whole unit of information. Yet for a language to be powerful and efficient, recombining different parts becomes desirable. So, finding subunits, or the smallest units of information, that might recombine is akin to looking at human phonemes and how they recombine to make words. Human phonemes are letters and vowel sounds that can be recombined to make new words. Take "fun" and "sun." If you look at a spectrogram, the phoneme "un" is basically the same in both words, but the beginning letters, *f* and *s*, look different, and we hear them differently, thus they are two different words.

This raises the question whether animals make sounds, or phonemes, that can be recombined to make different "words." Most scientists have always considered a "whistle" one unit without studying whether parts of a whistle could be recombined (a hallmark of language) to create different contours. In our team's first analysis, we looked at our clusters of sounds, and it looks like parts of whistles can be recombined. Of course, there are only so many ways a whistle goes up or down, but getting at the smallest unit of information will help us see if animals have anything like human phonemes. There may be two units that often combine and some that never combine with another sound.

Although we have always measured a whistle as a full unit of information, what if there is a little twist or turn at the end of a dolphin's whistle that represents a different signal? After all, Mandarin Chinese has a bunch of recombined contours that can mean different things depending on whether the end is up or down, as described by linguists Seth Wiener and Rory Turnbull.[3]

There are well-studied whistle languages. The famous French scientist René-Guy Busnel first described human whistle languages

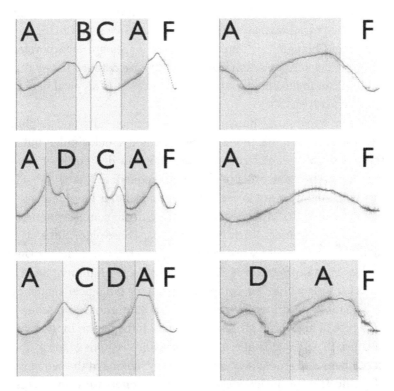

Figure 6.4. Searching for the units of information in a whistle, we can see how parts may be recombined to make different whistles. Recombination is critical in language since it allows an efficient system to be used. Courtesy of Daniel Kohlsdorf.

in the Pyrenees mountains, and his student Julian Meyer explored other locations with this phenomenon.[4] They asked the question, How much information can be encoded in a human whistle? The whistles encoded basic information, but these communication systems also contained a lot of contextual information. The whistling sheep herder was always on a particular mountain or whistling at a certain time of day, so there was a lot of extra information that the listener down in the valley heard. Although limited, human whistle languages do serve a restricted function and are an effective long-distance communication system, although the system is context sensitive (where the whistler is, the time of day, etc.). A few dolphin researchers have explored the content of these systems, but in the case of human whistles, the content was limited and dependent on the physical location and individual whistling. There are also human click languages, which use various acoustic parameter variations, as explored by linguists Wen-yu Chiang and Fang-mei Chiang.[5] Most human click languages are incorporated with other communication signals, but their existence shows that content can indeed be encoded by a diverse set of atypical human word structures.

When we compare human words, prairie dog calls, and dolphin squawks, for example, the forms of the sounds look similar, but is there really order and structure there? But how do I go about looking for structure and order in my sounds sequence? After all, that is my main goal if I want to see if any animal sounds have language-like structures. To find order and structure in our sound sequences, we turn to alignment-learning algorithms. These algorithms originally stemmed from work describing the human genome. One of the original programs is BLAST (Basic Local Alignment Search Tools), an algorithm for comparing primary biological sequence information, described by biologist Leda Cummings and others.[6] This type of algorithm gives us sequential information in the form of rules. BLAST is a database query that finds regions of local similarity between genetic sequences (e.g., by looking at the sequence of amino acids or nucleotides of DNA, researchers were able to discover different proteins, disease locations, etc.).

The basic idea is that you can have a stream of data types—let's say the four bases that make up DNA: adenine, guanine, thymine,

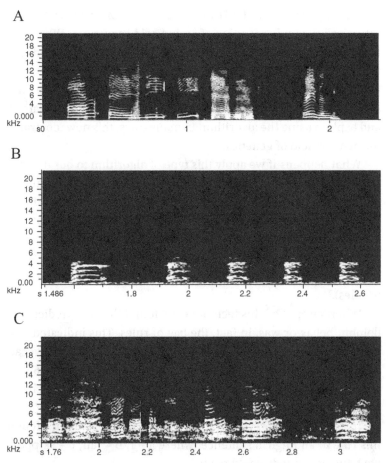

Figure 6.5a, 6.5b, and 6.5c. Human phonemes, *A*, look structurally like prairie dog alarm calls, *B*, which look structurally like dolphin squawks, *C*. Looking at the structure of human words, we can see that other animals create similar sound types. But finding out what they mean to an animal takes more exploration. Courtesy of the Wild Dolphin Project.

and cytosine—and the sequences can describe these smallest units of DNA. The sequences of different amino acids determine the composition of a specific protein. So, this algorithm gives you biological sequence information. The tool also allows genetic researchers to see where there are deletions or substitutions. Such changes might indicate a mutation or a different code for an area on the genome that might represent a tendency toward a disease.

The application of BLAST to genetic researchers' datasets revolutionized the understanding of amino acid sequences. When this tool became available to every researcher, the field exploded; the tool transformed the way we looked at and interpreted genetic sequences and their functions. The real success was because a platform was created that many scientists could upload their data to and explore using the algorithm. In many ways, this new tool transformed the field of genetics.

What happens if we apply this type of algorithm to our dolphin vocalizations? In fact, Daniel Kohlsdorf applied this technique to our dolphin vocalizations for his PhD dissertation. Daniel input one year of our dolphin vocalization data into a program that was essentially used for human speech. It extracted categorizes, including *bag of words* (individual units of sounds), *bag of* n-*grams* (clusters of sounds), and *bag of rules* (sequences of sounds that adhere to rules).[7]

When he applied this technique, he found the best predictor of dolphin behavior was, in fact, the bag of rules. This indicated that there was indeed structure to the sequences of the dolphin vocalizations. One of the rules we found can be seen in figure 6.6 (the bottom line). In this rule, there are some substitutions in the chain, showing how varied a rule can be.

Instead of dolphin vocalization types a, b, and so on, if I found this rule in an English sentence (where a = I, b = ran, c = skipped, etc.), the upper code could read:

> I (a) ran (b), I (a) thought and thought (d,d), I (a) happily happily (g,g) skipped (c)

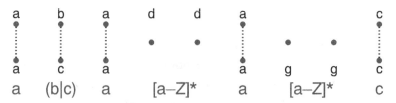

Figure 6.6. Rules represent structured aspects of sound sequences. Languages all follow some kind of order or rules. Here we see one rule that was found using dolphin data: sound *a* is followed by either sound *b* or *c*, which is followed by sound *a*, then any sound *a* to *Z*, and so on. Courtesy of Daniel Kohlsdorf.

Then the bottom line would read:

I (a) skipped (c), I (a) thought and thought (d,d), I (a) happily
   happily (g,g) skipped (c)

Kind of a boring example, but it gets the point across. Or a very simple rule might be "a" follows "b," or good morning. The point is that all language shows order and grammar.

These types of rules, that show substitutions and deletions, have been used very successfully in the study of amino acid chains. This is what alignment-learning algorithms can do: they can show you rules and patterns within your data. However, you could say that a pattern can be found in any large dataset, so the idea of finding clusters of words, or some order, should not be surprising. But to get at some of the elements of language-like recursive structure, we would expect to see something like a bag of rules. For example, with "Mike read a book to Fred," we would see "MRABTF," and I could inject this segment in this sentence: "I heard that MRABTF." Although often considered an aspect of human language, recursive structure is not always even obvious in normal human conversation. It turns out that humans really don't need recursive structure in their language; it just makes it a bit more efficient. So whether we should hold animal language to this criterion is debatable. It may make language more efficient, but we might want to reconsider it as a necessary quality for language definition. Nevertheless, we can look for structures-within-structures in animal vocal sequences. (See Arik Kershenbaum and colleagues for a comprehensive review of sequence analysis.[8]) After all, efficiency might also be important for other species' communication and for the evolution of animal language systems.

In my work with the dolphins, I am lucky because I have underwater video recordings to go along with my sound files, so I can begin to try to interpret these rules and patterns relative to the real world of the dolphins. In my software program, each segment gets a label. I have eight types of upswept whistles (labeled Ua to Uh), eight types of down-swept whistles (labeled Da to Dh), and eight types of burst pulses (labeled Ba to Bh). I might have the sequence the computer has recognized in a larger stream of vocalization types, and within this larger stream the computer has labeled a short sequence of sounds as "Uc, Dg, Uc, Dg." Let's say I have the same larger stream of vocalizations and the computer has noted a different sequence, "Ud, Db, Ud, Db," within the same larger stream of vocalizations. I happen to know that "Uc, Dg, Uc, Dg" is the signature whistle of a calf named Delphi, and "Ud, Db, Ud, Db" is the signature whistle of a different calf named Laguna. This discovery would essentially mean that two different dolphin mothers used the same sequence of sounds, changing only their calves' names in the sequence. This could be something like, "Pay attention, Delphi. I said come back," or "Pay attention, Laguna. I said come back."

Some other deciphering processes also exist. Many, like natural language processing, are well used in human language because we can check them with already-known words. Although very productive for human languages, these may or may not be great methods for animal language exploration.

A creative technique was described by Google team member Aren Jansen and his colleagues in 2017.[9] The idea here is that, at least in human text or speech, some words are usually found near other words. They are found in association with each other, so this is called nearest neighbor technique. This technique has been developed to use relational information to look for clusters of words. This led to the development of context-sensitive grammar, which assumes that words can be sensitive to the surrounding context or surrounding structure of a sentence. In fact, this technique is currently working in your Google searches; surrounding words or sentences

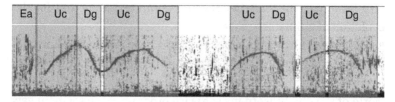

Figure 6.7. The UHURA (Unsupervised Harvesting and Utilization of Recognizable Acoustics) software program recognizes sound categories and labels spectrograms for the researcher to review. By mining data in this fashion, we can search for similar patterns in other sound sequences. Courtesy of the Wild Dolphin Project.

pop up as you search with a word or word stream. For example, I begin typing "how do I fix" into Google, and what pops up are various ways to finish my search phrase: "how do I fix my credit," "how do I fix a zipper," "how do I fix a leaky faucet," and so on. If I continue typing "how do I fix a broken," what pops up is "how do I fix a broken toe," "how do I fix a broken vase," and so on. Google search algorithms are using common context of things found in a search with the words "how do I fix a broken X." This is essentially context-sensitive grammar—my words are sensitive to their placement and surroundings of other words. It's a pretty good place to start for animal language-like structure exploration, because it suggests there is a logic in the communications. And like all human languages, a substituted word creates a whole new sentence and meaning.

There have been forays into applying many machine-learning techniques to animal vocalizations, including dolphin whistles, by researcher Mahdi Esfahanian and colleagues, to small ape vocalizations by researcher Dena Clink and colleagues, and to structural complexity by researcher Julia Fischer and colleagues.[10] And even though there are mathematical techniques, like information theory, that may tell us how complex a species' signals are, these techniques do not illuminate the structure of a language. Information theory doesn't interpret meaning or extract patterns, but it can tell us the relative complexity, and efficiency, of the communication system. Animal behaviorist Brenda McCowan and colleagues have shown that information theory can compare levels of complexity between communication systems, in these cases, with acoustic signals.[11] It turns out animal communication has long been thought to be a potential analogue for decoding alien signals, should we ever receive them. These same researchers have also analyzed the importance of Zipf's law, the measurement of the frequency of the most common words, when assessing other species, but this is more an illumination of the process of learning and word frequency than it is of language.[12] Convergence of the process and complexity in languages is interesting, but it does not suggest structure or grammar. Nor does it illuminate any details of information encoding or interpretation.

When information theory work first emerged, I found it interesting but lacking in depth since it didn't really get at the language itself, nor did it have any way to decode or interpret signals. Although we have a complete understanding of our human languages for the most part, the dolphin data used in these initial studies of dolphins only included whistles. From my over thirty years of experience with wild dolphins, I believe it is the burst-pulse sounds that have the greatest bandwidth and comprise the bulk of dolphins' social communication. Of course, humans continue to grab and analyze whistles because they are easy to measure and easy for us to categorize. And, realistically, that is the type of data most researchers have collected. I wonder what would happen if we included all the dolphin sound types into information theory. Even when we are trying to compare across species, we must remember that the quality and diversity of the dataset itself may be biasing our answer. Although information theory is a measure of complexity, this technique does nothing to show the details or meaning of a language.

How Do We Interpret the Meaning in Signals?

The biggest challenge in decoding animal communication will not be finding order and structure. It will be trying to interpret the meaning of such structures. This will involve the overlay of metadata, or extra all-encompassing data, to help with these interpretations. In dolphin work, this means knowing who was there, what their relationships are, and any other life-history details that might be relevant.

How do humans analyze meaning in a stream of written words or text? Clearly face-to-face interaction gives us multimodal cues of information. An anthropologist would, after observing a human culture, probably have an understanding of words and human behaviors. They would then enter and interact with the culture and engage individuals to fully understand the meaning of the culture's communication signals. But how do you interpret the meaning of communication signals within an animal society? Assuming we have categorized, clustered, and found rules in sequences of sound, how then do we interpret things? The answer is by using metadata, and later playback of vocalization sequences to test the results.

Metadata can mean a variety of things, but in the case of my own dolphin work, it means that when I collect a sequence of underwater behaviors and sounds, I know the individual dolphins present, their age, their relationships, and perhaps some of their history and personality. I can follow on the video, if I am lucky, who is doing what to whom, thereby creating a societal and individual context for the information use. This is analogous to an anthropologist recording a mother and child versus an adult male and female. The function and outcome of the communication need to be interpreted in the context of the society, as clearly described by anthropologists who have lived with other human cultures.

Of course, many anthropologists study nonhuman animals for various comparisons. And in the case of social mammals, we do know a few things. We know that there are critical developmental periods, that they need to eat and mate, and that many animals learn from or are taught by elders. So, in some ways we already understand some of the context of their lives. We might even understand etiquette and politics of a sort. Having a baseline dataset of an animal society, or the animals' situation, can help us immensely in interpretation.

There is good evidence that many mammal species experience the same types of pressures when raising offspring—keeping them safe, disciplining them, and so on. I happen to know that Atlantic spotted dolphins, like the well-studied bottlenose dolphins, have individual acoustic names, or signature whistles. In fact, signature whistles have been described since the 1960s, so no new news there. With enough examples of streams of behavior, and some metadata and knowledge of the species, I might be able to understand basically what is being communicated in a stream of vocalizations. But unlike when studying human societies, we might be limited in the sense that we can't "ask" a member of the animal society what is going on. Or can we?

Eventually, to really try to understand the meaning of animal communication signals, we may need to use either playback or an interface of some sort. Interfaces have their own challenges, as we have seen already, including both adequate technology and access to the animals we want to interact with. Nevertheless, we may need interfaces in the future to really test language-like structures and our theories behind them.

Now that we have learned about the process involved in machine learning and what it might give us, there are some new results from different animal studies that demonstrate the power of these techniques. There are many exciting animal communication projects. Biologist Livio Favaro and colleagues suggest that penguins have linguistic structure in their call sequences, researcher Stuart Watson and colleagues report that gorillas have structure in their vocalizations, and animal behaviorist Arik Kershenbaum, one of the leaders in applying machine learning to animal vocalizations, and colleagues report that canid howls vary across species and subspecies.[13] Most recently, biologist Taylor Hersh and colleagues analyzed the click patterns of sperm whales and suggest that they have a symbolic use, directly inferring language.[14]

Also exciting is the work by biologist Tim Sainburg and colleagues, who have used some machine-learning methods across species.[15] It is always very hard to compare across species. Their technique is a creative use of these new tools, and it allows us to begin to explore and compare across species to see if there are any universal rules. If nothing else, it gives us a comparative tool for measuring communication systems. We may discover both their similarities and their differences. I remember, after reading one of Tim's papers, I immediately contacted him at the Gentner Lab at University of California, San Diego, and discussed what techniques were most useful. Although they focused on birds in the lab, it turns out we were on a similar path, so the power of finding another researcher using similar methods, and finding interesting results, lends support to a potential powerful direction for exploring animal communication complexity in this new world of machine learning.

Of course, there are other types of hardware that help reveal the unseen world of animals. Hidden cameras and robots are great examples that have been used for both wildlife monitoring and animal behavior studies, and this technology offers a unique look at many difficult-to-study animals. BBC's *Spy in the Pod* series delighted us with a robotic turtle who caught the attention of and eventually infatuated a real turtle. And a robotic squid and its ultimate demise when discovered by a group of dolphins was entertaining. Marine mammalogists Sean Twiss and Joanna Franklin used robotic instruments to study the reaction of gray seals to novel objects.[16] Hidden cameras also promise to provide a glimpse at what animals do when a human is not trying to observe them. Although we all try to be benign observers, it's not easy. An animal may hear or smell you when you think they don't. And even if your study group is habituated to your presence, there still may be some unwanted effects of your presence.

For studying animal behavior, hidden sound-recording devices in the ocean have similar potential to hidden cameras in terrestrial environments. You will recall that in 2013, my team and I had just begun our field season when we came to realize that our dolphins were not anywhere to be found. We searched east, west, north, and south without any luck. We talked to local boat captains who also noted the lack of dolphins during their expeditions. Finally, we discovered all fifty-two missing dolphins, but they had moved a hundred miles south, across deep water, to another shallow sandbank. Since a few dolphins had not moved away, they would now be significantly harder to find in the very large ocean.

Like many other marine mammal researchers, we started placing recording devices so we might capture sounds while we were not present. Called passive acoustic monitoring (PAM), this technique involves using devices you can anchor to the ocean bottom or suspend in a water column for months at a time and then retrieve the data later. They essentially sample twenty-four seven when humans are not there. This technique gives you a collection of sounds, over time, that can be separated by species, time of day, and so on. In our case, we wanted to know when the few remaining dolphins passed by certain areas, with the hopes of timing our work when they were around.

We placed two devices in two separate locations on the sandbank. After one month, we picked up both units, downloaded the data, and then quickly analyzed it. We were then able to see a pattern of time of day and tidal state that predicted the presence of both spotted and bottlenose dolphins in our two different areas. The ability to categorize basic sound types (clicks, high-frequency whistles, etc.), and then translate them to specific species (Atlantic spotted dolphin or bottlenose dolphin), was invaluable. We were able to time our next trip to the exact tidal cycle, and sure enough, our dolphins were there.

Figure 6.8a and 6.8b. *Top*, Dr. Herzing holds a passive acoustic recording device (Ecological Acoustic Recorder, or EAR). *Bottom*, First Mate Tyler Hazelwood (*right*) and Wild Dolphin Project board member Drew Mayer (*left*) secure the EAR to the ocean bottom to collect sounds over one month. Photographs courtesy of the Wild Dolphin Project.

Of course, there have been all sorts of tags you could physically put on dolphins for many years. First there were radio tags that you could put on a whale (usually by mounting the tag using attachments that would pierce the whale's blubber layer). Radio tags have a limited distance of signal reception, so either you got lucky tracking a whale along the coast, or you could fly a plane and try to retrieve a signal, at about a distance of three miles or so. Then came satellite tags, which as you can imagine, retrieve their information from high-flying satellites, giving the researcher a broad view of an animal's movement. Although these tags are invaluable for tracking many endangered species, in our dolphin work, we have tried to be relatively noninvasive, as our purpose is to observe animals in the water. The only time we did satellite tag an animal was in 2018, the event discussed earlier, where Lamda, a stranded dolphin from our group, was rehabilitated for three months and then released. We worked with multiple researchers, including Charlotte Dunn and Diane Claridge of the Bahamas Marine Mammal Research Organisation, and released and tracked Lamda over a four-month period. He traveled down the western side of the sandbank, almost to Cuba. Finally, he returned to his home area and has remained there ever since (the tag fell off after about five months). A happy ending to the Lamda story, and a good example of the kind of information you can only get with a tagged animal.[17] What was also incredible was how Lamda reacted to us in the water his first summer back in Bimini. My research assistants would often remark on his friendliness, as he would rub up against them when they were holding the video camera. After working with humans for four months during his rehabilitation on land, I have no doubt that Lamda had a positive association with humans, and perhaps some gratitude. If you have ever rescued a pet, you know that they seem well aware that you have helped them.

There are also suction-cup tags that a researcher can put on a wild animal; they only last a couple days, but that is long enough to get short-term movement information. Marine mammalogist Robin Baird and colleagues have discovered many specific patterns of various whales and dolphins around the Hawaiian island chain using satellite or suction-cup tags.[18] So, depending on your question, there are many tools available. In my study area, young spotted dolphins remain difficult to identify each year, since they have limited spots or marks on their bodies. We always joke that we wish our dolphins had bar codes as they pass by our underwater cameras. It would save us hours of work trying to figure out who is who as they get new spots and marks on their bodies.

Although we don't have a Rosetta stone for what the dolphin sounds might mean, it does seem that we have some great tools and techniques to both gather and analyze animal communication signals. You might ask why we don't have a Rosetta stone for animal communication. As described in this book, there are many facets to understanding another species. In the past that understanding has been difficult without a large dataset, metadata, and the complicated tools that are only now emerging. Yet AI tools like machine learning offer great hope for processing and understanding other species, given the right situations. Some are tools we could only dream of decades ago. Some are tools that really will allow us to ask simple questions about the structure of animal sounds, like whether units of sound get recombined or if there is a grammar or order in a sequence of sounds. With the right, rich dataset, we may be able to overlay patterns we find with the metadata about the society and try to interpret the meaning of the sequences of sound. If possible, we may need to use playback to test the function of a sequence of sounds to really do a final test on how good our interpretation is.

In the next chapter, we will take a closer look at the definitions and evidence for both language-like structures perhaps already discovered in some species and the strict definitions of language that humans hold tightly to their chests. Have animals already reached out to us? How have we interpreted their actions? Can they use the tools, words, and interfaces we give them to communicate with us or each other differently? Do animals really have language-like structures in their own communication systems? Or are they just exposed to language by interacting with humans, thus propelling them into a new, but enculturated, world of human communication?

7

Every Species Has Its Ambassador

An individual can move a society, or its concepts, forward—even individuals of another species.

We have already seen throughout this book how species understand each other in the wild, including a variety of taxa. We also see that, first and foremost, most species need to prioritize the encoding of information about danger or types of predators or hazards—likely because this has great survival value. Humans also need to recognize danger and to communicate specific dangers to each other. We also communicate to find mates and food and, as with other social species, to establish friendships. Understanding communication systems is critical to the study of all animal behavior. Interactions between individuals are often mediated by vocal signals. These signals play a role in determining the outcome of both intra- and interspecies competition and other social interaction.

Scientists over the decades have worked with a variety of species (primates, birds, dolphins) using experimental methods to probe their minds and cognition. Keep in mind that animals exposed to humans, or human communication systems, become quite adept at observing and learning our system of communication. The watched become the watchers. You could say they have been enculturated, human enculturated, in a sense. At a minimum, they have a heads-up on how we communicate, and they can incorporate this into their interactions and potentially into their manipulation of a situation. Remember, comprehension of a language comes before production, even in human children. Your kids are listening and learning long before they can talk. It may be that, like human children, animals who are exposed to communication signals may comprehend words before they can use them functionally. What, then, does this suggest to us if we are to explore decoding or even interacting with another species? And what have we learned so far from work with other species?

Feathers, Fins, and Other Animal Ambassadors

Much of what we have learned comes from some seminal studies over the last four or more decades. As we have seen, much of it has come from interactive interfaces between humans and other species. Most of this work has been based on work with a small subset of individuals of different species. We must remember that even in our human society, not all brains are created equal, nor are personalities across individuals the same. So, the idea that certain animals emerge as "Einsteins" should not come as a surprise to us. These individuals, who seem eager to engage with humans, are often the ones who become ambassadors of their species. Are they more intelligent? Or do they just have more curious personalities? We should be careful not to generalize a species based on the results of a couple of animals. We wouldn't do this in a human study, and we shouldn't do this in an animal study. But we must also recognize that working with large numbers of animals is not always easy or feasible. And in some ways, the "Einsteins" of the animal world emerge, whether from boredom or desire to please humans, naturally during these studies. Some emerge as ambassadors for their species, eager and willing to work with humans on tasks and experiments. Others just don't have the personality traits—they may be shy or scared of humans, unconducive to such work. Or it may be that an animal just isn't interested in interacting with humans, either because of its personality or due to a past negative experience.

We have plenty of scientific studies that show personality

variation in different taxa, such as horses, dogs, octopuses, and dolphins.[1] My own graduate student, Nate Skrzypczak, analyzed some of the wild Atlantic spotted dolphins in the Bahamas using different personality measures, including sociability, curiosity, and boldness, based on the percentage of time individuals spent with conspecifics, human researchers, or their mothers.[2] There was a strong connection between boldness and curiosity, suggesting a strong personality correlation. Skrzypczak used this baseline of personality measures to predict neophiliac (curiosity toward novelty) behavior specific to human-dolphin interaction. Such predictions both verified and projected best dolphin candidates for some of our more interactive fieldwork. Bottlenose dolphin personalities in the wild have also been studied by researcher Bruno Díaz López, who showed that personality also mattered within wild bottlenose dolphin communities.[3] So, humans are not the only ones with varied personalities driving our culture.

Personality can have a large impact on our measure of intelligence and cognition in any species. And now we know that dolphins have important and functional variation in their personalities, and this variation helps make up a healthy dolphin society. Much of the early animal research focused on primates for obvious reasons. Humans are primates, so many of the body signals and postures of other primates are recognizable to us. Working with nonprimate intelligence is a different matter. As we have already seen, and hopefully learned, we can be blinded by our senses, or lack thereof, and biased by our idea of what intelligence is and how we might recognize it. While we have been successful on many levels, real breakthroughs have been made with only a few species. In some cases, the failures have been as illuminating as the successes, leading us down new paths and providing awareness of our own biases. In the first primate studies, when we tried to get chimpanzees to learn English, our initial failure to recognize that chimpanzees simply could not anatomically create human words eventually led us to explore their use of hand gestures, a more natural communication modality for chimpanzees, albeit more challenging for humans to decipher. We also tried to get dolphins to speak human words, and again, we discovered that dolphins just don't have the anatomy to make the same types of sounds as humans.

The individual nature of most interspecies communication studies is also striking. Alex, the African gray parrot, was a star pupil of animal behaviorist Irene Pepperberg.[4] Kanzi, the bonobo chimpanzee, emerged as a star student of language by observing his not-so-successful mother work on a keyboard with researcher Sue Savage-Rumbaugh and colleagues.[5] And Akeakami, a bottlenose dolphin in Hawai'i, learned to comprehend artificial languages successfully with the consistent work of psychologist Louis Herman and colleagues.[6] But were these individual savants? Did they have similar types of personalities, or did their human proctors specifically mentor them somehow to be successful? If we measured their personalities, would they all be bold, initiating types? Or would some be shy, only to emerge when humans are in the picture?

Recent cultural work on other species suggests that individuality is key for a healthy, complex society. Zoologist David Lusseau has reported the detailed roles that individual dolphins might play in maintaining a larger societal network.[7] Some individual dolphins are networkers and others are information brokers. What happens when a networker is lost? Do the connections between groups break down? How does the work of individual animals affect the survival of a species in the wild? Could it be time to stop thinking about animal societies just as numbers and instead understand that diversity of individuals might be equally, if not more, important?

Historically, it is important to remember that most of our work with animals has been done in captive situations or using experimental protocols, and most of it has been one-way experiments performed while exploring the cognitive abilities of other species based on human definitions and under human control. Our pets, who we interact with most regularly, can show a sophisticated understanding of human words and moods, although communication is rarely really two-way. As we have already discussed, ingenious types of interfaces and techniques have been used over the decades to learn about the minds of animals. So what have we learned?

First, we have learned that animal interfaces benefit from social interactions that include humans. Communication is social, and technology on its own is boring. A machine can give you food and objects, but only another being can give you an interaction that relates to your biological nature and social needs.

We have also learned that animals that are exposed to human words, on a regular and rich basis, can learn to comprehend the function of many words. However, most interfaces are one way, so they lack a two-way system for the animal to communicate back to the human. Without an interface, some animals have learned to combine and create words to attempt to communicate a concept (as we saw when Koko used the American Sign Language signs for "sweet" and "water" to describe watermelon).

We now know that a species may or may not need a technological interface, depending on their sensory systems. Some animals require interfaces to communicate with humans because of the disparity between sensory systems (e.g., dolphins are acoustic, bees are ultraviolet). These interfaces need to be species specific. Some species can easily mimic human words, so they already have the tools to communicate back to us (e.g., birds that mimic words). Over the decades, a variety of unique technologies have been designed and employed to access the sensory systems and cognitive tendencies of different species. For example, primate work has focused on manual dexterity tasks and computerized "touch" keyboards, and bird work has tapped into species' vocal mimicry abilities. Adapting such work for dolphins requires special technological considerations due to their aquatic environment and their acoustic sense, which is often primary to vision.

Do Universal Features of Communication Exist?

Some of our direct experiences with animals might suggest that there are some universal features that cross species boundaries. I believe the most probable candidates are the prosodic features of communication, which likely transcend culturally agreed upon words or structures. These features include intensity, frequency, and duration of a signal. You can have an intense vocal sound, an intense postural signal, or an intense (dense) chemical signal. It is the same with frequency (number of units of something in time) and duration. You can have a high-frequency tone or a high-frequency postural signal (rapid shaking of the head), and you can

have long drawn-out words or short rapid words. But do these prosodic features transcend senses and thereby species?

From a biological and evolutionary perspective, is there enough
overlap between species to make it possible to communicate with
the help of prosodic features? If so, communicate on what level? Or
do differences in our sensory systems and communication signals
make it forever impossible to have a chat with another species?
Back to biologist Eugene Morton, who discovered the universal
features of acoustic communication signals within mammals and
birds, and who suggested that there is continuity in the evolution
of at least some acoustic communication signals.[8] Remember that
Morton showed that high-frequency, narrowband sounds and low-
frequency, broadband sounds have a similar function with both
birds and mammals. Perhaps all these prosodic elements evolved
first, in the process of communication, with words and language
coming afterward. We know that a spoken human language often
has cadence, a variety of intonations, that enhance and supplement
meaning. Does your voice get louder and lower when you repri-
mand your dog? I know mine does. At a minimum, your voice signals your emotion: you are not happy that your dog ate your shoe.
Duration and amplitude can often encode emotion, no matter who
we are talking to.

In a study of interspecific signals used between humans and
domesticated dogs around the world, animal behaviorists Patricia
McConnell and Jeffrey Baylis found that dog trainers and handlers
have all independently evolved similar acoustic signals to communicate with their dogs.[9] These signals evolved separately across
different places in the world and time periods. This suggests that
there is an inherent biological aspect that helps shape these signals, in this case between species. Other studies between humans
and dogs and horses by biologists Jean McKinely and Thomas Sambrook have also shown understanding between species.[10]

Such independent development can be considered an aspect
of convergent evolution, meaning that a feature can evolve independently from the same pressures in the world. I am always reminded that the shape of a dolphin is the same shape as a shark
since they both need to travel through the water as hydrodynamically as possible. If, in fact, we consider the possibility that ani-

mal and human communication systems may incorporate similar features, regardless of the sensory modality (i.e., visual, acoustic, chemical), then we might consider these features as important if we are looking for universal aspects of communication. Researcher Piera Filippi has similarly noted that more and more we are finding that prosody is a potential feature that may cross species boundaries.[11]

It turns out that our sensory systems are designed to translate a certain amount of information across our senses; this is called synesthesia, which refers to the crossing of the senses and is well studied in humans by researcher Lawrence Marks.[12] It turns out that we are neurologically wired to perceive one sense and translate it through another. Most humans can touch an apple blindfolded and then identify it visually after being unblindfolded. Dolphins can echolocate on an object underwater and then identify it visually as well. It's a handy skill to have, and it likely helps in the learning process. Then there are extreme examples of synesthesia, often found in artists. Mozart was synesthetic and could see music as colors and shapes. Some folks might hear notes and see color. Ordinary people taste specific flavors when they hear specific musical notes. Or they might liken an emotional feeling with a specific color. Or a musical note or chord may have a sharp edge to it visually. Do animals also have overlapping sensory systems that might accommodate bridging the gap by using nonvocal features such as rhythm, timing, and intensity of communication signals? Perhaps we will discover some universals that transcend the medium and create the same signal across species.

Could these prosodic and emotional cues be the trick for engagement with another species? Is this the key to looking for universal aspects in interspecies communication?

We know these features are quite important between humans. Remember how the tone of a conversation can be lost in an email or text message; the importance of expression has heralded the creation of emojis for our electronic communications. It might not be the most frequent signal that is exchanged, but it is one of the most salient. Mutually created signals, or emotionally powerful signals based on previous interactions, are perhaps the most effective during interspecies interactions and communication. This suggests that we can cocreate the meaning of signals when we interact with other beings. Sometimes called "situational awareness," or in cognitive science this is also termed "embedded cognition or distributed cognition," this is the idea that you mutually create communication through, and give meaning to, actions and responses.

In past animal communication studies, we have not been willing to consider that animals might be intelligent and able to communicate complex things. But scientists have not really had the tools to analyze complex behaviors. New frameworks are emerging, including dynamic systems theory (DST) and distributed cognition (DCog), which promise new and complex tools for the analysis of communication. In her book *The Dynamic Dance*, anthropologist Barbara King describes the framework and specific examples of how communication signals are produced and negotiated in real time.[13] In DST or DCog, the process of communicating socially is not about transfer of information but rather the emergence of mutual understanding through a shared action or thought. Meaning does not reside in words but in the mutual construction of meaning between partners, created through interaction and coregulated by all participants. Increased coordination is seen between individuals as they reach meaning through negotiation. Such mutual adjustments might be seen in chorusing male chimpanzees adjusting their pant hoots in the process of call convergence. Dolphins adjust their physical movements and their vocalizations to reach physical and vocal synchrony during aggressive chases. Even with exciting new frameworks to explore, it is most important to let the data, not current theory, guide the research and point to the most promising directions.

Meaning can exist during the interactions of two separate species if they engage as social partners. Interspecies interactions exist in many forms, and most don't include humans. Close interaction depends on close attention to detail, knowledge of the other, and subtle adjustments in interaction. Observing unfolding events, to another species, may just be the normal process for their social world. Play is an interspecies media, an opportunity to engage and explore the other. In two species with bodily similarities, such as spotted dolphins and bottlenose dolphins in the Bahamas, the markers may be closer and more recognizable than two disparate species.

The repertoire and use of communication signals in nonhuman species may be more complicated than we have imagined and might have evolved to converge on several things—many may prove to be translatable across species. Temple Grandin, well known for her research on both animal behavior and autism, can see and appreciate the world in ways that others cannot, and she learned to empathize with animals because of her senses and sensitivities.[14] Grandin discovered many practical ways of helping the livestock industry to reduce the stress of their herds. She discovered simple ways to change an animal's demeanor and keep the animal calm. It's their world, not ours. And Grandin was able to see and feel things that others could not.

Nonverbal communication signals cannot be underestimated in their contribution of both passive understanding and active communication in most species. In a study of crows using human gaze and facial recognition, researcher Barbara Clucas and colleagues showed that American crows were able to use human gaze as a visual cue, giving them information about human intention.[15] Furthermore, in an experiment involving mask-wearing humans who did risky and dangerous things, researcher Bernd Heinrich and colleagues described how social learning in ravens spread knowledge about these "dangerous mask-wearing humans," showing the resilience of interspecies interactions and perceptions of intention.[16]

Animals are far more aware and complex in their communications than we have been willing to give them credit for. Since one of the only criteria left for separating human intelligence from the rest of the animals is language, can we think of other ways to measure intelligence that are more species specific and less human biased? After all, a human is not evolved for the water, so our abilities to see at night underwater are inadequate, whereas a dolphin can see somewhat at night and can use echolocation to hunt or flee as needed. We would simply be shark bait. So, really, every animal is intelligent in its own environment and has been driven through evolution to excel in its milieu.

Intimate interspecies social interactions, or observations, often raise the question of self-awareness. How self-aware are dolphins? How could they not be aware of their own selves with such complex and intimate interactions with other dolphin species? How do you explain some of these interactions without speculating that the dolphins are using planning or language?

Regarding measuring human intelligence, we have discovered that the old IQ test has had many biases over the years, such as cultural biases, and that it did not measure humans adequately. Developmental psychologist Howard Gardner has listed "seven types of human intelligence," including emotional and musical.[17] So, we might come to acknowledge other "types of intelligence" in nonhuman species someday.

I have been involved in the astrobiology community for decades. Examining nonhuman minds on Earth often leads to questions explored by scientific writers for decades: If you discovered intelligent life in space, what would it look like? How would you recognize it? The concept of convergent evolution—that similar features can emerge independently, like large brains or eyes—has always fascinated me. And I wondered if we might find features and measures of intelligence that have also evolved independently across species on Earth, thereby suggesting that they could evolve elsewhere in the universe. So, in 2014 I decided to do a little exercise, with the help of some colleagues, to measure nonhuman intelligence on a variety of traits. I labeled this multidimensional exercise to profile and assess different types of intelligence as COMPLEX (COmplexity of Markers for Profiling Life in EXobiology).[18]

Using a structure similar to the Myers-Briggs psychology typing in four dimensions, I listed the following five dimensions of intelligence to score for each taxon: EQ, or encephalization quotient, essentially a measure of brain-to-body ratio; CS, or communication signal complexity; IC, or individual complexity, like personality; SC, or social complexity; and II, or interspecies interaction, the tendencies to engage other species. Each dimension had four attributes to score for each species. Then I enlisted a few experts/ colleagues who worked with these nonhumans: dolphins, social mammals; bees, with collective intelligence; octopuses, solitary animals that use tools; microbes, rapidly evolving and diverse organisms; and machines, programmed by humans but displaying "intelligent" behavior. I asked them to score the nonhuman's attributes in each category, based on what they knew from the critters they studied and from the known scientific literature. For example, experts in EQ scored brain-to-body ratio, neural density, neural specialization, and convolution. An expert in IC scored personality, role of individual, leadership, and flexibility. Experts in CS scored for sensory modalities, natural repertoires, information theory, and symbolic information. An expert in SC scored for a measure of group living, alliances and cooperation, network variation, and culture or social learning. And II experts scored for natural interactions, cross-species altruism, sensory gap to humans, and evidence of enculturation. It was a very small exercise and not a full-blown study, but I wanted to explore how we might begin to look across different dimensions and measure nonhuman intelligence.

A multidimensional overlapping view allows each animal to be seen where its most powerful potential exists, in this case relative to other species. If we look at a subset of three examples, a dolphin, an octopus, and a bee, we see that both the dolphin and the bee scored high on complex communication signals, but the bee has a smaller brain, and the octopus is clearly not social. How do we want to interact with these species? Do we want to try to communicate with them? Or just leave them be? This can be a pretty good way to think about the abilities of a species.

In the original paper, you can see all the species tallied, and you will note the differences, that a dolphin scored high to medium in all five dimensions, an octopus scored high in brain (1) and personality (5) but low in interaction (3) and sociality (4). Bees scored low in brain, interaction, and personality but high in communication (their waggle dance is symbolic and highly specific) and sociality. Microbes scored high in sociality and interaction. Machines skewed toward brain, communication, and sociality. Although insects score high on communication, they don't seem to show much social interspecies interaction. Octopuses are also very intelligent and can use tools, but they show hardly any social interaction. Of course, dolphins, social mammals, excel in social and even interspecies interactions. It was pointed out to me, after my presentation at a conference, that microbes are so diverse (and include bacteria, viruses, etc.) that it would be hard to generalize. But that really is the point. Within each group of animals, there are many species and many diverse attributes. And we have already noted

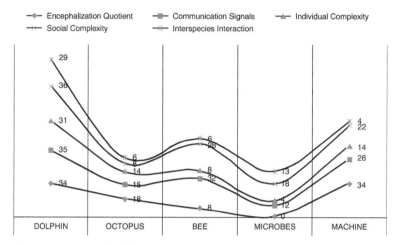

Figure 7.1. Comparing features across taxa, or species, can help illuminate different types of intelligence and desirable aspects when engaging other species. For example, a bee may show a low score for encephalization quotient (brain-to-body ratio) but would score high (32) on its use of communication signals, which have aspects of language. A microbe scores higher on interspecies interaction than an octopus because of its tendency to interact with other species, whereas an octopus is solitary most of the time. Courtesy of the Wild Dolphin Project.

that personality of individuals may also vary, although this may be more of a social mammal quality, but personalities have been found in fish and other nonmammals. So, there is an enormous task to be done in the future if we really want to build a tool for cross-species analysis for intelligence. Developing this type of exercise could be very helpful in the future to help us avoid past mistakes, such as trying to compare self-recognition in mirrors by different types of animals, like apes, who are visual, and dolphins, who are primarily acoustic. As humans, we are biased, preferring a visual test since we understand this sense better.

Of course, we have done a horrible job even acknowledging other species on this planet and their right to exist and not to be disturbed. Because we have technological developments to destroy their habitats and claim them for our own development and species use, we have ashamedly disavowed their rights. This human behavior may be an ethical issue and moral issue as much as an intelligence one. And humans still do horrible things to each other, like go to war for territory, lie for money, and deceive on a regular basis. I have often struggled to understand this objectional part of human nature. But because we do have the ability to invade and inhabit more of our environment, we just do—often we do so without considering the consequences to other species or to the planet itself. It may be our ultimate downfall. I suspect it comes from a deep disconnect with the idea that we are part of nature, not above it, and this disconnect has destructive consequence to the natural world.

If you begin your studies with the idea that "humans are unique," you will never be able to see the full capacity in animals. Cetaceans have gained status and recognition as having cultures, as described by researchers including Hal Whitehead and others.[19] In fact, the loss of specific groups of animals can be considered something analogous to genocide, as distinct animal cultures are wiped out. So, when humans passionately hold tight to their use of language as a uniquely human ability, it may be our last grasp at human superiority—after all, we can kill at will, so we feel dominant. Language allows us to communicate about the past and future; it allows exchanges of abstract ideas and feelings. It also allows for complicated manipulation and deception. Language is probably one of the great determinants of our survival since the Ice Age. Can we imagine that animals, having evolved sometimes for millions of years, could have evolved complex ways of communicating that either rival human language or, at a minimum, incorporate some language-like structures that satisfy their needs?

In an interesting analysis of interspecies science, animal cognitive researcher Irene Pepperberg asked the question, What happened to all the animal language studies?[20] After her review of the history and the trials and tribulations of decades or research, Pepperberg noted that scientists have missed many opportunities to explore convergent evolution across species because of the lack of support for follow-up research.

What, exactly, are the criteria for language as we know it today? Remembering that communication can take many forms besides acoustic, including visual (sign language, chromatophores, and postures and gestures), how do we compare what we have found to the human definition of language requirements? Humans have many criteria we think need to be included in a language. There are a few criteria that we will need to address if we are to answer the question of the existence of animal language, keeping in mind that "absence of evidence is not evidence of absence."

Language entails many features, including symbol use (words), time displacement (past, present, future), combinatorial aspects (recombining vowels, consonants), and other features yet to be found in their totality in other species. In human communication, we see evidence of the crossing-over of senses, called synesthesia, which opens a fascinating avenue for animal communication research. Do animals also have overlapping sensory systems, or universal features in their communication system, that might accommodate bridging the gap between them and humans, using nonvocal features such as rhythm, timing, and intensity of communication signals?

Things We Already Know

First, we know that many *animals show cognitive flexibility*. We know that some species at least have the ability to understand artificial languages, so their brains have the structure and flexibility to understand another species' language. That doesn't mean they have their own language. But if they did, how would we know? How might they show us? We do have cases that involve other species trying to communicate with humans, so how do they try? In the end, they try to communicate with what they have available, either their natural abilities to create and modulate information or an interface or acquired "language" that humans have given them.

Second, we see that many *animals use tools*. Tool use has been included in one of the main definitions of intelligence for decades. We know that many animals use tools, for a variety of reasons, and in a variety of ways. Chimpanzees use sticks to dig out termites, ravens use sticks, otters use rocks, dolphins use sponges to poke at poisonous fish and octopuses, and other marine organisms use shells for shelter. So, the word is already in on this aspect: other animals use tools.

Third, many animals show some *complexity of information in their repertoires*. Although not fully studied across species, due to the challenge of acquiring a full repertoire of signals, the repertoires of some species have been measured using information theory and Zipf's law (both showing us levels of complexity, with human language as the comparative bar). Although these methods show complexity, they do not illuminate things like grammar or language.

Things We Don't Know

First, we don't know if animals have any *structure, order,* or *grammar* in their repertoires. This may be answered soon with the number of researchers now using machine-learning tools and the like. With these new tools and techniques, we should rapidly be able to determine whether there are rules and grammar in animal systems. We know that every human language has some sort of order or grammar, so that information would be a tremendous start. Then we can begin to overlay all that we know about their societies (the metadata) to further decipher the meaning of patterns. Recursiveness (using a phrase within a phrase), somehow got on the list of traits needed, but humans don't really need or use recursiveness in functional language use. Why, then, should we try to search for it, or expect it, in animals?

Second, we need to know if animals are able to communicate *time displacement* and use *referential labels* in their repertoires. Although we do have evidence for word labels or referential communication (e.g., alarm calls for types of predators) in both vervet monkeys and prairie dogs, we have yet to look at many other species. The additional benefit of labeling things is that you can then potentially talk about things in the past or plans in the future. Maybe you are eating at the dinner table, but you are talking about what happened in school today. In fact, this is one of the major criteria of language: that you can communicate displaced in time and space. You can talk about things from yesterday or in the future. In some ways, referential communication is necessary for time displacement signals. After all, if you grunt or whine at the dinner table, it probably reflects the real-time feelings. Yet coupled with a referential label/words, it might express how you felt yesterday about an event or person. Personally, I believe this is one of the most powerful aspects to look for in an animal communication system. It seems like it would have great survival value: "Meet me at that great hunting ground so we can eat well," or "Beware of that younger male. He is very aggressive with newborns." Nothing would be more valuable in real-time planning, or even reflection of relationships and life. And with so many species, at least social species, having great memories and long-term relationships, it raises the question whether these species might be capable of using time displacement and referential labels.

If I were a mother dolphin, I would certainly find it helpful to communicate to my inexperienced little calf that the shark I just sensed on the bottom is a dangerous bull shark and not a harmless nurse shark. This would be a signal that would rate high in survival value through evolution, much like the alarm calls that label specific predators for vervet monkeys or prairie dogs. The convergent evolution of such signals in aquatic mammals should not really come as a surprise since the need to survive probably drives the evolution of such signals on land and in the water. However, we have yet to show such signals with dolphins, but that may merely be a limit of our abilities to gather data in the large ocean environment. I know over the decades we have seen various shark interactions with dolphins, and although we are in one of the safest places on the planet to work in the water, we still find ourselves getting out of the water when a fifteen-foot tiger shark approaches. You are still in the food chain. So, data collecting is not always easy in these circumstances.

Finally, the notion of finding *universal features of communication* could help greatly in the process of studying interspecies communication. As we have already seen, Eugene Morton's research has shown that there are both structural and motivational rules, based on the harshness and the frequency of sound that many birds and mammals follow.[21] Patricia McConnell's research on the similarities of whistles used by humans to communicate with dogs has shown cross-cultural patterns that naturally emerge.[22]

Of course, many of us believe that language is the most complicated product of intelligence. So, if you want to answer the question about nonhumans having it, you must be willing to search for it and apply new tools to perhaps previous data. You must be not only willing but able to do so, and that involves time and money. So not only do we need visionary scientists—we also need visionary supporters. Such science is considered high risk. But many MacArthur scholars and entrepreneurs acknowledge that you must be willing to risk new ideas and work to make progress and push a field. We have seen this with satellite missions to detect exoplanets. And biologist Barbara McClintock's creative discovery of the jumping corn gene won her a Nobel Prize.[23] Luckily science funders and benefactors have recently been willing to consider funding high-risk science money well spent. So, again, we may need to look to our machine-learning tools to help us parse signals and interpret them with larger metadata. Although some mathematical programs (information theory) have suggested the possibility of other complex communication systems, showing language-like structures and features will require very specific datasets and algorithms. This will be a long process, both in the acquisition of the data and in processing it in a creative and illuminating way. The final challenge will be to interpret and understand any structures we find, and that will likely happen by playing back specific signals and sequences of vocalizations in the water or air to the very animals we study to see how they react.

It turns out that some of the most progressive thinkers, and very progressive foundations, have recently begun to support this complex area of study. My organization, the Wild Dolphin Project, benefited from a three-year grant for this work from the Templeton World Charity Foundation through its Diverse Intelligences initiative. Designed to explore the myriad of minds and intelligence in nature and in the world, everything from animal emotions to the basis of consciousness itself, this dynamic and visionary program's mission is to seek out cutting-edge research and provide the space and time to put these thoughts and tools to work to explore the presence and diversity of animal intelligence. This foundation promotes the exploration of "big ideas" and "big questions." In the next chapter, we will look at not only the obvious big questions but also other peripheral big questions and areas of study that might be impacted by the discovery of an animal language.

8

Big Claims Take Big Evidence

Make me a liar, Fish.
—*Contact* (1997 film)

What are the implications of interspecies interactions on Earth and beyond Earth? The emerging field of astrobiology has shed new light not only on organisms on Earth living in extreme environments but also on the escalating count of planets in our galaxy that may potentially hold life. How will we approach a new life-form on another world? Are our intentions clear on what we will do if we find intelligent, or at least sentient, life? Will work on interspecies interactions on our planet give us some practice interacting on levels, and with minds, previously unknown?

The challenge is to understand what makes sense in the exploration of the world and minds of other species without biasing our conclusions—or doing other species harm. Can we take some of our cognitive questions out into the wild, where animals have evolved their real-world skills? Along with our exploration of cognitive interfaces in the wild will come the challenge to redefine our definitions and measures of intelligence.

Can We Develop a Definition of Intelligence That Is Not Species Biased?

Intelligence is a tricky thing to measure even in human subjects. Clearly every species could be considered "intelligent" in its own world. We should be looking to develop species-specific definitions for "types" of intelligence, rather than resorting to human comparisons. Researchers have attempted to measure the intelligence of animals in various ways. There are encouraging findings that suggest that the communication signals of a species inform us about intelligence in terms of complexity, information, and structure. Scientists have always struggled with the definition of human intelligence, and we have had all sorts of biases emerge from the IQ test, for example. Some are cultural differences, some are gender differences, and so forth. So how, then, do we measure animal intelligence?

How useful this typology is might be a matter of our intent toward other species. Do we want to just measure types of intelligence a species uses in the ecosystem? Do we want to interact with a species? If so, on what scale? Should we be merging decoding and interaction elements when looking at other species? I can imagine a tool, say a real-time translator between humans and animals. We essentially have this type of tool across many human languages already, mostly due to Google's work on translation, both written and verbal. Keep in mind that this takes millions of examples of words and their context to translate. So, the datasets, either photos for visual or conversations for audio, have contributed to the amazing rise of translation options.

What can we learn from a really different, but very successful, type of intelligence, such as a swarm-type intelligence? We might want to understand the mechanisms of coordination, but do we want to harness this intelligence? Do we just want to understand it and how it fits into the larger world? Or do we want to communicate with it, to engage in mutually beneficial tasks? In the early years of working with chimpanzees, specifically with sign language, biologist Roger Fouts speculated that we might be able to go into the wild with our laboratory signing chimpanzee and have the chimpanzee translate what is going on with wild chimpanzees.[1]

It's an interesting idea, but at the start I can see one big problem. If you have a captive chimpanzee, one born in a zoo or brought to a zoo at an early age, it really hasn't been exposed to normal chimpanzee culture, and likely not to all social signals. There is also a dolphin analogue to this idea. While I was visiting a dolphin facility in Curaçao many years ago, one of the trainers described to me their ocean-going program, where they trained bottlenose dolphins they had captured from the wild in Honduras to go out to the local reef and swim with divers. The wild-caught dolphins did fine. But if you took dolphins who were born in this facility out to the reef, they were afraid. They didn't know what a reef was, the sounds were probably strange, and they probably didn't know how to catch a fish. They simply did not grow up with the sounds and tastes of the sea and therefore feared it. Dolphins and chimpanzees need "street knowledge" to survive in the world, and captive animals will never have this since they are housed and fed by humans and thus dependent on humans for survival. Although they might not lose a lot of their instincts, the animals are essentially domesticated, or at least handicapped.

Another way we measure intelligence is by looking at physical infrastructure. This includes the brain-to-body ratio, termed encephalization quotient (EQ). Neuroscientist Lori Marino places dolphins second only to humans on this scale, even above the great apes.[2] As we have seen, researchers also measure intelligence with cognitive tests, including language comprehension, problem-solving abilities, and mirror self-recognition tests. For example, psychologist Louis Herman and colleagues reported that their bottlenose dolphins in Hawai'i could comprehend both gestural and acoustic language, including word order (syntax), word meaning (semantics), and abstract concepts (two dimensional images on an underwater TV as well) over decades of research.[3] Chimpanzee researcher Sue Savage-Rumbaugh has tested for problem-solving abilities and cooperative communication through shared tasks.[4] Another aspect considered critical to advanced intelligence is mirror recognition, the ability see yourself in the mirror and recognize the image as you and not another of your species. This has been documented in primates by researcher Gordon Gallup, in dolphins by neuroscientist Lori Marino and behaviorist Diana Reiss, and in elephants by researcher Joshua Plotnik and colleagues.[5] Complex counting problems as well as memory and other cognitive skills have been documented in various birds (such as crows and ravens), dolphins, and domestic dogs, as reviewed by researcher Michael Beran and colleagues.[6] So convergent evolution of physical and cognitive abilities is manifest in many intelligent species since complex societies, both on land and in the water, deal with the same "types" of problems in their social worlds, from politics and complex communication signals to long-term relationships.

Dolphins are obviously nonterrestrial in their design; they don't have hands or feet to manipulate objects. So what are the dolphins doing with all that brainpower? It is likely that they spend much of their time negotiating social relationships, developing friendships, and monitoring the environment. Understanding the evolution of how communication systems are learned and transmitted may help us envision the breadth of possibilities. Some ingenious techniques for exploring the process of shaping language have been explored by scientist Olga Fehér; they could be quite productive when we think about different types of intelligence and useful in the process of recognizing intelligence in different forms.[7] Emotional expressions have been noted by scientist Beáta Korcsok and colleagues in nonverbal communication.[8] Biologist Raphaela Heesen and colleagues have described linguistic laws in the gestures of chimpanzees.[9] Play may even have implications for development and interspecies interactions, as noted by cognitive researchers Heidi Lyn, Patricia Greenfield, and Sue Savage-Rumbaugh.[10] Even the segmentation of vocalization units in animal sounds may be related to the evolution of human language, as discussed by researchers Dan Mann and Marisa Hoeschele.[11]

The last decade of animal communication work has seen progress in acquiring large datasets and the application of both traditional statistics and new algorithms and tools to help categorize and understand signals. Signals can be measured for a variety of parameters (e.g., frequency, amplitude, duration) and by repertoire aspects, including the size of the repertoire (call types), the complexity of information encoded (information theory), or the structure or grammar of a system. Communication signals have evolved in different physical environments where certain mediums of transmission may be favored (e.g., acoustic signals travel well in water, visual signals in air, electromagnetic or optical signals in space). Social environments, such as complex mammalian social systems or insect communities, may also drive the complexity of signals. We know that different signal types can be modulated in similar ways. They can be frequency modulated or amplitude modulated, and their durations can be modulated. They can be modulated relative to one another in a time series; they can be modulated to emulate, mimic, or synchronize with other signals. They can be discrete signals or graded. Their information can be encoded in their modulation of the signal directly, or it can be encoded symbolically. So, these important signals can be single or take on complex structure and order. We clearly need increasingly creative methods for fully assessing communication systems and their "types" of intelligence on a broad scale. More and more scientists are finding ways of measuring intelligence, or defining types of intelligence and cognition, that are species specific rather than human driven.

We have also found that dolphins can "perceive the other" in significant ways. One example from my work in the Bahamas illustrates this. In one encounter, dolphins showed their recognition of the human situation during the disciplining of their young. Calves, excitable as they are, sometimes get out of control, even when tended by a responsible babysitter. Although young calves may not understand safety boundaries at this age, it is possible that they are testing their boundaries with the adults. During one swim, Katy, a young female spotted dolphin babysitter, came by me and swished and slapped her tail in front of my face as if to say, "You're getting these calves too excited, and it's my job to control the situation." I quickly stopped paying attention to the calves. Later, a mottled male, Big Wave, chased a postpartum female he wanted to mate with while people unknowingly interrupted his attempts. Big Wave swam by me, swiping his tail near my face to communicate, "Yield to me my space to perform my social function." Without trying to be anthropomorphic, that is the message I got. Although it could be interpreted as aggression by an inexperienced eye, to me it was clearly the transmission of their social rules and an assessment of their, and our, behavior. I believe this is how they communicate rules in their own society. It is interesting that the dolphins often address these signals to me, possibly because I am always in the water, and it may appear that I supervise these other humans to a certain extent (which I do). That would mean they recognize roles and responsibilities in another species and direct their communications appropriately. My colleague Dr. Thomas White and I wrote a paper in 1998 about these scenarios and how they add to the list of qualities of "personhood" for other species.[12] It is a sophisticated skill to say the least.

New Perspectives

NASA researchers landed the rover *Perseverance* on Mars for a ten-year mission to look for evidence of life, or past life. Humans are living on the international space station and looking through the eyes of the Hubble Space Telescope and the new James Webb Space Telescope—all the way back to the big bang and the creation of the universe. We are finding the elements of potential evidence of life on other planets, including water and amino acids; thousands of exoplanets have been identified, and on Earth, bacteria live in extreme environments where, until recently, we were unaware life thrived. What will happen when we stretch our boundaries and meet a sentient or pre-sentient life-form elsewhere in our universe? Interspecies dialog, a true dialog, with another sentient life on Earth may show us a way. Of course, we have already encountered other intelligence on our planet, and our record is not good. We use, exploit, compete with, and exterminate—because we have the power we dominate. What if we changed our definition of power? What if we considered empowering other cultures, other species, other habitats, and our immune systems? What might that future look like?

As we start exploring habitable exoplanets in our galaxy, the study of nonhuman communication signals is often suggested as an analogy for discovering and decoding extraterrestrial signals. I have been involved with SETI (Search for Extraterrestrial Intelligence) for many decades. My interest overlaps with this broader search for life in the galaxy. I believe that—especially now, when we have discovered thousands of exoplanets, some habitable by our definition—we will eventually meet other species in space. And how will that go? It might be determined by our very ethic developed by studying species on Earth. Our earthly interactions can be an analogue for extraterrestrial life elsewhere. Of course, most biologists agree that we will first find microscopic life, or at least the elements of life, before any complex life-forms.

Remember that SETI is really the search for civilizations, specifically technical civilizations that can transmit radio signals or optical signals. It is quite as likely that nontechnical cultures exist elsewhere, as they do on Earth, given the many social or swarm species that do just fine without technical abilities. But SETI's search rightly focuses on long-distance communication, listening at first for radio signals. Recently SETI has expanded to include a search for far-traveling optical signals (lights, lasers) and technosignatures (any other technical signal or artifact showing a civilization). There is also a program called METI (Messaging Extraterrestrial Intelligence) led by researcher Doug Vakoch.[13] METI, the active version of SETI, has been controversial. Broadcasting our location in space with messages from random individuals or facilities may be dangerous. The late Stephen Hawking considered it a mistake for the future of the human race. If a nonbenevolent civilization with greater technological abilities hears us, Hawking wondered whether humanity and planet Earth could be in jeopardy.[14] Of course, many a sci-fi movie has been made about invading aliens. So actively sending out messages remains controversial within the radio astronomy community and the larger SETI community.

I have mixed feelings about METI, since I both passively observe and intentionally interact with the wild dolphin community I study. Although I patiently waited for decades to understand their society before interacting, even small mistakes along the way were potentially harmful. In 1991, when I played back the dolphins' signature whistles without fully understanding the etiquette about which individuals make whistles in certain situations, I learned a valuable lesson. I didn't repeat this experiment until much later, when I had better exposure to and understanding of the dolphin society, after having worked with it for decades. After years of observations, we finally decided to create a dolphin-friendly tool to empower the dolphins to interact with us on a more detailed level. I have no doubt that one of the reasons that we haven't had much progress using interactive communication tools with dolphins is that the available interfaces simply have not been easy for dolphins to use. Like chimpanzees, dolphins don't have the ability to produce human words. But unlike chimpanzees, they don't have primate-like appendages to touch a keyboard. They have their rostrums (beaks), and in captivity they have been taught to use them to trigger a keyboard. But wild dolphins do not normally touch strange objects. Instead, they use their acoustic abilities to investigate them.

Recall that, in 1997, two top-level dolphin cognitive psychologists, Adam Pack and Fabienne Delfour, joined me in the field to explore the use of an underwater interface with our community of spotted dolphins in the Bahamas. I had met Adam while visiting the Kewalo Basin Marine Mammal Lab in Honolulu, and I met Fabienne when she worked at the late researcher Ken Marten's lab at Sea Life Park Hawai'i. Ken and his colleagues had created an underwater computer for the dolphins.[15] The dolphins could "point" to the computer screen by orienting to it through an underwater window, which triggered (by breaking infrared beams) the computer inside. Simultaneously, John Gory and colleagues were building an underwater keyboard for dolphins and humans at the EPCOT Center in Orlando.[16] This first-of-a-kind underwater keyboard was a great model for what we wanted to do in the wild: provide a keyboard accessible to both dolphins and humans.

The Delfour and Marten experiment yielded some interesting insights into the importance of modeling and interaction. Their interface allowed the dolphins to select a circle, a square, or another object on the screen by pointing to the area on the screen, which effectively broke an infrared beam (the same idea was used in the EPCOT keyboard). Dolphins would subsequently hear sounds or see motion on the screen. The computer was a breakthrough for being a good interface for the dolphins and for giving us a chance to learn what the dolphins enjoyed doing. Fabienne and I quickly saw that the dolphins soon got bored with the computer screen. As we observed their interaction with the computer, Fabienne and I reached the same conclusion about how to make progress with this technology. Based on the breakthrough work that Irene Pepperberg had done with Alex the African gray parrot, we believed that the problem was that the dolphins needed a model, a demon-stration of how the system worked and why they should want to engage with it. And there also had to be some social impetus for them to use an interface.

Remember that one of the very important contributions by Irene Pepperberg had been reviving, with tremendous success, an old modeling technique called "model/rival." Pepperberg designed experiments using a human as an equal participant with the parrot, Alex. Once Alex accepted this human as the equivalent of another parrot, Pepperberg used her interactions with that human partici-pant to demonstrate to Alex what words and concepts meant. She engaged Alex in a pseudonatural situation, competing for social attention with the other "parrot." Her work revolutionized ideas about how other species think and express their thoughts, partic-ularly because she was working with a nonprimate, rather than a chimpanzee.

As discussed previously, by adapting the model/rivalry tech-nique to the wild, we were on our way to creating a two-way in-terface to communicate with dolphins. We allowed the dolphin participants in our study to observe humans using the communica-tion system in the water and receiving rewards for certain behav-iors. We hoped and expected that the dolphins would then model this behavior and compete to use the underwater keyboard, with the human participants as well as with each other. Incorporating new technology and basing our work on everything we knew about dolphin communication and cognition, Adam, Fabienne, and I set out to crack the code of communication in dolphins.

One of my biggest heartbreaks occurred during the develop-ment of our two-way work with the wild dolphins of the Bahamas. We had been working for over thirty years monitoring and describ-ing the life and communication system of this aquatic society. It was only after we began our two-way work, between humans and dolphins, that the media really pounced on our research. Although our dolphin research had already been covered by the BBC, *Nature*, *NOVA*, and others, it seemed that most humans were only interested in dolphins when it was about connecting with humans and inter-acting. As someone who had been fascinated and frankly happy to just observe them and understand their own society, I found this both shocking and sad. Really? We had been meticulously gathering data on the dolphins and working on deciphering the dolphins' communication for decades, yet to many folks it only mattered that we were exploring how to incorporate our findings to the advantage and entertainment of humans. My motive for even exploring a two-way system was because of the dolphins and their habit of reaching out to us in the water in many intense ways. Perhaps, I thought, intelligence seeks intelligence, and part of our process is to be human ambassadors, showing the dolphins what is possible when two species meet each other with respect. Of course, we were the minority in the human world, and I often worried that the dolphins, like other wild animals habituated by humans, might someday run across some not-so-nice humans and be taken advantage of or worse, abused and harassed. But we carried on with our work, assuming that an intelligent animal would have some discretion and ability to sense the intentions of humans.

But despite our good intentions and creative technology development over the years, in 2020 our work, like many other scientific projects, was interrupted by both COVID-19 and climate change issues. Although we continue to work on decoding our vast database of sounds and behaviors, our ability to collect new data is challenging and limited by human activities around health and advocacy. Around the globe, pollution, new oil drilling activities, and environmental degradation devastate and disrupt many of nature's patterns and support structures. So, while the scientific endeavor does go on, humanity continues to ignore the shouts and cries from the natural world.

I think about the astronomical tools that revealed the number of new planets in our galaxy and reshaped our thoughts and our ideas of our place. In the same way, new tools for deciphering and understanding animals will surely shape our worldview.

Acknowledgment of the Other

If other species meet the "language" criterion of intelligence, what will we do? Will we argue that because they aren't like us, they still have no rights—because they can't build buildings (well, beavers do) or develop mathematical formulas (or maybe they do but just use them instead of writing them down)? Do we have the capability to act morally and ethically toward species we don't understand? In his book *In Defense of Dolphins*, ethicist Thomas White argues for the rights of dolphins based on many of our own criteria for personhood.[17] Humans, after all, are not the only species on the planet, but we often act as if we are the only species that matters, and this has severe consequences for other species, ecosystems, and the planet itself. White argues that a species has the right to flourish, meaning they have the right to life, healthy environments, and freedom of choice.

Even at the SETI conferences, the debates heat up over why there is no evidence of other intelligence signals from outer space yet. The argument by some goes, "Well, if they are out there, where are they, and why haven't they contact us yet?" Again, such a reaction supposes that humans are the center of the universe and so interesting that no species outside our own solar system could pass us up. Presuming that there are thousands, if not millions, of other species or civilizations out there, we may be of no consequence except to ourselves.

So, what are the implications of studying a nonprimate intelligent species? The deciphering of an animal language could be one of our greatest feats. What is more intriguing—the differences or the sameness? Will new insights about animal intelligence make us more distinctive, or bring us into a larger community of beings, of nature? The possibilities for communicating with another species abound on this planet, and dolphins top the list despite their nonprimate appearance. It may yet be our best training ground for exploring the cosmos for other life, for if we can't understand and interact with life on this planet, then there is little hope for our exploration of the galaxy. The day will surely come when we will meet non-Earth-based life-forms, microscopic or macroscopic, and how we face dealing with them is up to us. New research groups are springing up to try to study whale communication. Gašper Beguš and colleagues have recently claimed to have found the equivalent of "vowels" in sperm whale clicks.[18] At the Earth Species Project, researchers seek to catalog and analyze many different species using the newest and best AI tools.[19] Recently Brenda McCowan and her team in Alaska have produced evidence of mimicry between a humpback whale and humans.[20] And my team and I continue our working deciphering spotted dolphins sounds with our machine-learning interface and analyzing the previously described mimics between humans and spotted dolphins.

Human beings are selfish and tenacious about their uniqueness. Not that we aren't unique. By any standards, humans have unique qualities, as do most species. Can we develop nonbiased species methods that allow us to explore, with respect and mutual participation, the lives of other species? It's a bit like the prime directive in *Star Trek*, which demands that explorers not interfere with the normal development of any species they encounter. Visionary producer Gene Roddenberry points to a future where we recognize that interfering with another society or culture has great consequences—to the culture, to ourselves, and to the evolution of the universe. I think, perhaps, it is simply inconvenient to acknowledge the intelligence and complexity of another species. How much are we willing to give up and to change to allow other species to exist in a healthy way on this planet? We continue to expand human populations at an alarming rate, mow down rainforests, and pollute the ocean as if we are the only species that matters.

As we emerge from our slumber and begin to listen and make sense of other species' communication systems, we can begin to use such knowledge to further explore evolution and universal communication issues.

Endless Possibilities

The possibility of communicating with animals has fascinated humans as early as Aristotle. Interspecies interaction, specifically human-dolphin interaction, is a very old phenomenon but only recently a field of scientific inquiry. Stories about dolphins befriending people continue to be reported, including dolphins rescuing swimmers, guiding lost vessels, and aiding people who fish in their catch, as recently updated by researcher Paulo C. Simões-Lopes and colleagues.[21] Biologist Marc Bekoff, in his book *The Emotional Lives of Animals*, explores the possibility that, like humans, animals may suffer from psychological problems like depression, autism, and post-traumatic stress disorder.[22] Does the elephant that stomps and kills a handler, or an orca that grabs and drowns a trainer, have memories of being beaten, as we know too well happens to animals, potentially driving such behavior. Perhaps the feisty dolphin in a swim program remembers witnessing the slaughter of her family while dolphin trainers grabbed her away to sell into a program thousands of miles away. How can we restrict a large mind, in sterile and constricting environments, without understanding these implications? The very qualities that make these animals curious and social are the same qualities that may cause them mental harm.

The issues and concerns raised by studying interspecies interaction are fascinating and important—some are species specific, others are ethical. There are methodological challenges and breakthroughs in technology on the horizon. Taking cognition and communication studies into the wild is now possible with our mobile technologies and high-speed computers. We can now use technology to bridge the gap between humans and animals, as we have bridged the gap between human languages.

Etiquette and ethics of first contact are important in establishing trust and intention. As they say, first impressions matter, and it is no different between humans and other species. From the groundbreaking and insightful 1992 book *The Inevitable Bond*, edited by Hank Davis and Dianne Balfour, we are reminded that historically both scientists and animals acknowledge and use bonds for mutual understanding.[23]

One of my own first lessons in "dolphin etiquette" occurred during my anxious attempt to determine the sex of a dolphin during my early years in the Bahamas. It was the middle of the summer, and I was in the water with a couple spotted dolphins. As I dove down underneath one of the dolphins to determine its sex, another dolphin moved in and placed himself in between the dolphin and me. They both swam off, rubbing pectoral flippers as they went. Later I would come to understand that the action of inverted swimming underneath a dolphin can be aggressive signal or an invitation to have sex. This was my first experience with dolphin etiquette and one I did not forget. After this mistake, I learned to be more aware of what messages I, as a human, was sending in the water. After all, we were In Their World, on Their Terms (our motto at the Wild Dolphin Project) and that meant learning and observing the local customs.

There was clearly dolphin etiquette to learn with wild dolphins. When I caught up with a traveling group of dolphins, they would open a space for me and allow me to move along with them. First, the dolphins "positioned" me accordingly in their group, but when I tried to change my position, as in the case of trying to get a photograph, it was met with jittery turns, glances, and basically breaking up of the dolphin formation. Other times I observed dolphin etiquette, including mother-calf and juvenile-juvenile pairs pec-to-pec flipper rubbing excitedly as they met up, or when they observed something novel in the water. Sometimes a trio of dolphins would engage in a pec-to-head rubbing event. One adult, slightly below and behind the other two dolphins, was receiving a pec rub on his head from each dolphin to his side. Trio pec rubs often occurred after fighting as potential peacemaking gestures, possibly analogous to chimpanzees reconciling by grooming. This clear and purposeful signaling system involved in pectoral-flipper rubbing related to the relationship of the dolphins involved as well as the body part they rubbed.

My first summer in the Bahamas had a very powerful impact on me—a feeling of being a student in the dolphin's classroom—a classroom of experience and experimentation. The key to this study would be to "be" in relationship with the dolphins: with respect, clarity, empathy, and a critically observational eye for whatever they chose to show me. If they were willing to invest in a relationship, or at least tolerate my presence, then I was willing to invest a few decades into nonintrusive observations. My hope was to blend in harmlessly and observe their lives with each other, not as an intruder but as a friend somewhat familiar with their culture. Forty years later this investment in trust and observation has paid off, as my team and I now observe our fourth generation of dolphins in the wild.

As in most research settings, a keen eye is required to ask the most detailed questions. A prime example of such a keen observation is the discovery of estrogen-mimicking chemicals found in the coating of test tubes during cancer experiments. Scientists couldn't figure out why the controls were growing cancer cells until they realized that the manufacturer of the test tubes had changed the coating, which turned out to be a chemical that encouraged cancer growth. This is why basic science is so important—because the process of discovery is valuable, in and of itself, perhaps leading to an answer to a yet unaddressed question. Today, funding comes strongly from the applied science side, "designate a problem, and find a solution."

Scientists should seek to illuminate natural laws, processes, and patterns. In his book *The Structure of Scientific Revolutions*, philosopher Thomas Kuhn describes the resistance of scientists to overwhelming data, and its implications, in emerging and controversial fields.[24] This is particularly relevant today with strong evidence of animal intelligence and emotions.

Many scientists, including me, are aghast at the lack of natural history data and observations today. I sometimes see this lack of engagement even with my own students. The process of spending time with the natural world is in jeopardy, causing us to lose sight of the natural world while at the same time we are trying to measure it.

In 1991 my old mentor, Ken Norris, visited us on our boat during our seasonal work with the dolphins. Joined by *National Geographic* photographer Flip Nicklin, Ken was writing an article titled "Dolphins in Crisis." Traveling around the world, Ken was bearing witness to not only the variety of dolphin species but also the variety of impacts humans had on the habitat of cetaceans worldwide. Flip joined us for six weeks to get some working shots and to document our research. He photographed a full-body image of me in the water with my video gear, and I jokingly told my relatives, after the article was published in *National Geographic* in 1992, that I was a centerfold, with the crease of the magazine cutting through my now globally exposed thighs.

Ken in many ways was the father of dolphin research. As a student in San Francisco, I eagerly drove down to Santa Cruz to listen to Ken lecture in his marine mammal class. Originally a herpetologist, Ken was an old-school naturalist, and he inspired many a graduate student. Fellow students Jan Ostman and Michael Poole went off to Hawai'i and Tahiti, respectively, to work with spinner dolphins, and I went to the Bahamas to work with spotted dolphins.

Ken was adamant about getting a look at dolphins' underwater world, and he had built a boat with a sturdy plastic tube that projected below far enough for a researcher in Hawai'i to sit in it and watch spinner dolphins. Known as the seasick machine, it provided mostly a behind-the-group view, as I was told later by his students. So, it was not a surprise how shocked Ken was when he joined us in the clear blue waters of the Bahamas and got a look at spotted dolphin behavior underwater. Having never seen this species, he remarked how different they were from spinner dolphins. "You would never see a spinner dolphin away from the group," he remarked, as we watched a rambunctious juvenile spotted dolphin play around the boat.

Later that night, drifting along the deepwater edge with glittering stars overhead, Ken told us the story of spinner dolphins in Hawai'i. He talked about the volcanic islands and the habitat this provided, about the rhythm of the spinner's day, offshore at night to feed and sheltering in the shallow bays during the day to rest. He was a storyteller, and a good one. His keen eye for natural history, combined with a strong sense of checking back with nature after data collection, were qualities I always valued and tried to incorporate into my own work. I often wished I could share with my graduate students the wonders of fieldwork as well as Ken did. Today, students come to collect data, forgetting to look up at the stars and think about the pulse of the animals they are studying and their environment. I think it is a loss for the world, to reduce animals to data instead of telling their story. As long-term mentor and colleague Scott McVay once told me during my early years trying to describe the underwater behavior of the dolphins, "It may not be as much about describing the behavior, as understanding the process of communication." I think these profound words hold true now more than ever, and that our work propels us to understand the laws of nature, as much as the details. Scientists can, and should, do more to study the laws and dynamics of the communication process.

I have never regretted taking time to learn the dolphins' rules at the expense of data collection. It was the best investment I made in the research to ensure its continuity. We were constantly tested on dolphin social etiquette—if we passed the tests, more secrets were revealed; if not, it was back to square one. Engaging the dolphins as full participants in the work, and in a personal relationship, has been my primary approach over the years, because without the dolphin's cooperation, as partners in this learning, we were nowhere. Our use of benign methods, like gathering fecal material for DNA versus capturing, tagging, or taking skin plugs biopsies for genetic work, preserved the unique opportunity at this field site, which is first and foremost to observe the dolphins underwater.

Dolphins belong in the wild. An average spotted dolphin travels at least ten to twenty miles a day, has a large network of friends and family, and invests years, if not decades, in teaching young dolphins how to survive. Dolphins teach their young complex skills, including feeding, babysitting, and negotiating the fine lines of dolphin behavior. A dolphin can only reach its true potential as a healthy, fully actuated individual in the wild. So, then, how can we give dolphins in captivity the ability to live out their lives in dignity with their own kind in a stimulating environment? There have been multiple failed attempts at creating dolphin rehabilitation and retirement facilities, usually stemming from human ego and politics. With enough money and the right intention, it is conceivable that we could create a dolphin retirement center. Models for successful retirement facilities for other social and sentient species, including primates (https://www.centerforgreatapes.org/) and elephants (www.elephantsanctuary.com), exist and are successful options to an undignified death or cruelty by isolation. A dolphin sanctuary could relieve the burden for facilities of dolphins destined to swim alone in small tanks until they die in the back room, unnoticed and forgotten. It could also provide a place for research, or education, based on the personality profiles and needs of the dolphins. Dolphin birth control could be provided if the goal of the facility would not be to create more dolphins for shows but to let these individuals live out their lives with respect. Some dolphins might choose not to interact with humans and others might step up to be ambassadors of creative work and help us understand the mind of a dolphin. Either way, it could be a win-win situation as long as the dolphins, not profit, remain the priority. Ever focused and committed to the ethics of animals, Lori Marino has recently spearheaded the Whale Sanctuary Project (https://whalesanctuaryproject.org/).

Created to establish long-term and realistic options for retiring cetaceans, current plans include an orca sanctuary and beluga whale sanctuary. Planned by multitalented staff and advisors (I am honored to be a scientific advisor), this sanctuary will prove to be a long-awaited option for captive dolphins and whales—a place to retire with dignity and peace. A way of making reparation the best we can for these animals.

I have always wondered what we are teaching children who observe captive dolphins. Are we modeling the wrong ethic? Even though I'm not anticaptivity (at least not until there is a retirement center available for existing dolphins), I am vehemently anticapture. Taking a dolphin or whale out of its pod is kidnapping by any standard. By displaying dolphins and whales for profit, we teach children (and adults) that it's okay for these animals to be used as entertainment, for people to hold a dolphin's dorsal fin and be towed around in the water or demand that dolphins and whales jump and leap on command—or, worse, that it's okay for people to ride a dolphin or orca in a vulgar display of human dominance. We are modeling to our children that it's okay to keep animals in captivity—large, intelligent animals with long-term memories, long-term relationships, and complex minds. Of course, we have done this for decades to elephants, big cats, and other complex creatures. Many circuses have now shut down their animal shows after realizing the public many no longer tolerate such abuse. Successful sanctuaries exist for many species, including elephants, rhinos, giraffes, and even farm animals. But until a retirement or rehabilitation center opens (which will be a long, expensive process) for dolphins and whales, and as long as consumers demand entertainment and therapy from dolphins, there will be captivity in many forms.

Many scientists believe that it is the challenge of living in a complex social group that drives the need for intelligence. One must have long-term memory to remember individuals within the society and their habits, and one must communicate effectively to negotiate and resolve conflicts. The ability to learn and pass on knowledge, specific knowledge unique to a local culture, could be advantageous. How might we begin to recognize other types of intelligence in nonprimate species? It is easiest to recognize types of primate intelligence, primarily because we recognize similar facial expressions and body language given our common ancestry and evolutionary trajectory. However, convergent evolution, the theory that species under similar selective pressures reach the same endpoint, leads us to think about other, nonprimate, species. There is evidence of physical continuity between species, like the hydrodynamically efficient streamlined shape of sharks and dolphins. Evidence of mental continuity may be found in species faced with parallel social politics, individual personalities, and group living, such as dolphins, elephants, birds, and, of course, a variety of primates.

I suspect we will also find emotional continuity in many species. Using functional MRIs, researcher Atilla Andics and colleagues discovered that, when listening to human words, dogs can separate word meaning and intonation, suggesting that this ability has evolved in nonhuman species, even in the absence of language.[25] Horses have been shown to understand and remember positive or negative interactions with humans, as described by biologist Serenella d'Ingeo and colleagues, suggesting that memory is used to predict and measure future interspecies interactions.[26] Usually, evolution and natural selection favor those traits that are adaptive. It stands to reason that continuity in intelligence or emotional expressions may be adaptive and something commonly found in the natural world.

Studying any species in its own environment is challenging. How is intelligence used in the wild? For example, we do know that most small dolphin species have complex social structures and communication, although their "language" is yet to be deciphered. While most researchers have focused on acoustics, language can also be of a nonverbal nature, such as sign language for humans, which encodes complex ideas and exchanges. We tend to only recognize language like ours, but we have to diversify our concept and analysis of language to really explore intelligence in other species. Cognitively, dolphins are both flexible and creative. Yet, the process of enculturation (by human contact) and exposure (to other communication systems) can be contributing factors to successful experiments in captivity. Researchers of primates have suggested that although some primates are able to perform complicated tasks, it is the exposure and embedding into a human so-cial context that encourages the emergence of intelligent actions. Could this be the case with dolphins? Do dolphins, like primates, have the structure and flexibility to learn but only show the emer-gence of these features when put into the right contextual or cul-ture circumstances?

The sci-fi book *Speaker for the Dead* proposes a scenario of a galaxy full of intelligent beings, some who are intelligent on their own and some who "might be intelligent . . . but we cannot know it." These are the species that the protagonist, Ender, is in search of, but he fails to recognize the signs of potential intelligence and instead wipes out a species that was what he was looking for all along.[27]

A dolphin, a being so different and alien to primates, can be taught artificial communication systems. But we still don't know if dolphins, in their normal communication system, refer to objects in their environment, to abstract concepts, or to past and future. We don't know if they can combine small vocal elements into large vocal elements, or if they have a systematic grammar that can pro-duce an infinite combination of meanings.

Can we truly look at a definition of intelligence in a nonbiased, nonhuman-centric way? To be socially intelligent, a species would probably have to be social, have a wide range of sounds/signals (and have control over them), and use language as a subset of com-munication with the ability to communicate abstract ideas. During my career in behavioral biology, which spans forty years, remark-able discoveries have been made that illuminate the complexity and intelligence of many species.

When I was in graduate school in San Francisco, there was a bottlenose dolphin named Spock at the local marine park. When he retrieved "garbage" from his tank, such as paper and floating objects that drifted in or were thrown in by visitors, Spock was re-warded by trainers with a tasty fish. One day a trainer noticed that Spock was bringing up rubber strips to gain his reward. It turned out that Spock had been tearing off the rubber caulking around the windows in his tank and was keeping a stash of rubber "currency," which he would trade at the surface when he wanted a fish. Spock had learned the system well and used it to his advantage. Perhaps these types of observations will prove the most telling of what another creature's mind is like and how it operates. Equally important may be the observations of "spontaneous" behavior, which can signal to us the creative or emergent features of an intelligent system. Other species may use spontaneous signals to enhance the clarity of communication in challenging communication scenarios. I suspect that most animals are more intelligent than we realize and communicate in more complicated ways than we know. We do know that other species can communicate sophisticated information to each other. In some species, the information may even qualify as symbolic (like the waggle dance of the bee).

The acceptance of interspecies interaction as a valid and important scientific area of inquiry will be dependent on the acceptance of something larger: a nonanthropomorphic-centered relationship with nature. Expanding our ethical boundaries to include nonhuman nature, and viewing ourselves as participants in life, will determine much of our future interaction with the world. We have an ethical and a moral responsibility to not destroy the complexity of life on Earth, or elsewhere.

The development of a participatory science is already underway, specifically in the areas of cognitive ethology and interspecies communication reviewed here. Like with other paradigms, boundaries must be crossed and a new way of thinking about possible methodologies and experimentation in science must emerge. Viewing nonhuman subjects as mutual, full participants in discovery is likely to increase our understanding of other species, much as cultural anthropology has attempted to do with different human cultures.

Historically, the evolution of ethics has come from the inclusion of the "Other"—other races, genders, and species—into a moral community, usually by proof of sentience. The emerging paradigm that allows exploration of interspecies interaction is larger than the current scientific one. It is not a paradigm about the potential intelligence and sentience of one species. It is about our appropriate interaction with the living world, of "being in relation with," rather than in control over. The great, late biologist E. O. Wilson, writing about biophilia, concluded that "reverence for life" would one day be understood in terms of evolutionary biology and evolutionary

psychology.[28] Perhaps this will give us new insight into the cooperative and interactive nature of all biological systems. Wilson's
concept of putting humans back on a parallel of psychological and
biological evolution with the rest of nature is related to our own
health, the health of our planet, and of our being.

In the book *The Tribe of Tiger*, author Elizabeth Marshall Thomas
describes the change in relationship between humans and a community of African lions over her lifetime.[29] The lions were used to
sharing a water hole with the human inhabitants of a village. After
farming became fashionable, a fence was put up restricting access
to the water. The lions could no longer easily get to the water. Humans and lions, who were respectful and unafraid before, within
a generation had become cautious and afraid of each other. Marshall's story speaks of the importance of continued attention to etiquette and ethics in relation to other species.

We are connected with everything, through our ecological cy-
cles, through evolution, and through our mental continuity with
other species. To immerse ourselves and be allowed to participate
in the world of other animals may give us access to the "*Umwelt*," or
details, of their world.

Sentience (normally defined by humans) does not guarantee
acceptance into a moral community or rights to integrity of life.
Paradigms limit the study of interspecies communication by making it an unacceptable area of inquiry, while retaining the anthro-
pocentric attitude that once kept the center of the solar system on
Earth. Replacing "being in control of" with "being in relationship
with" will certainly have an impact on our future. We can remain
in the old paradigm of being separate from nature and in control,
or we can acknowledge the many levels of interaction we have with
nature, including biological, psychological, emotional, and spiri-
tual; interspecies interaction is only one possibility. Crossing these
boundaries brings about deep cultural change. Our human desire
to connect with and contact other species, on a social level, may be
our future salvation. And from the conservation perspective, it's not
always about saving a species—it's about saving a unique culture.

If we really want to go down the road of interspecies communication, there are a few promising directions we could explore with
dolphins and other species. Most importantly, *a species must want*

to communicate. Familiarity with, and acknowledgment of, the importance of social bonds to the development of a mutual language cannot be underestimated. Decades ago, naturalist Ken Norris compared observing animal cultures with work done by cultural anthropologists Margaret Mead and Gregory Bateson.[30] He speculated that animal societies have rituals and ceremonies that help transmit and sustain the culture. To date, four species of cetaceans have been described by biologist Hal Whitehead and colleagues—killer whales, bottlenose dolphins, humpback whales, and sperm whales—that show aspects that favor social processes conducive to culture.[31] My own observations suggest that not only do spotted dolphins have complex mimicry abilities but they also have culture.[32] My guess is that most small dolphin species have culture, but they have yet to be systematically analyzed. And do the dolphins in the Bahamas have a culture of interacting with humans? That could be a new kind of culture. But disturbing the natural lives of dolphins by assuming they would like to interact with us is as dangerous as capturing them out of the wild. It does not acknowledge their presence as individual beings with lives of their own and the integrity to make choices about when and with whom to interact. I believe it is impossible to truly understand the essence of dolphins unless our participation with them is mutually voluntary and respectful. I think people are drawn to dolphins because they represent a healthy and live connection with Earth, and one that has the potential to re-create our place and relationship with nature.

Acknowledgments

In many ways I have been lucky to have developed a passion so early in life. I am continually grateful for my ongoing work with the Wild Dolphin Project, its staff, board of directors, and scientific colleagues. My administrative assistant, Melissa Infante, keeps the project running, while Britinni Hill, Hayley Knapp, Liah McPherson, and other field assistants keep the research going in the wild. Captain Peter Roberts painstakingly whips our research vessel into shape—lovingly, bolt by bolt—keeping us afloat during the field season. Newly minted board members Axel Stepan, Drew Mayer, and Nicole Mader, along with other board members, bring fresh new ideas into the work, while retaining In Their World, on Their Terms as a motto.

I am thankful to the Templeton World Charity Foundation's Diverse Intelligences initiative, whose exploration of the "big ideas" has helped many of us explore our data and our ideas about other minds. Bringing scientific minds together to explore nonhuman minds is a feat undertaken only recently by this cutting-edge program. Its support has allowed us the time to explore and develop our machine-learning work to begin to look for language-like structures in dolphin vocalization patterns. And much thanks to the many other foundations and individuals that have supported this work over the years.

My colleagues at the Georgia Institute of Technology have persevered in the development of much of our technology for machine learning and for the two-way communication system with the dolphins. Thanks to Thad Starner, Scott Gilliland, Chad Ramey, Celeste Mason, Daniel Kohlsdorf, and the many students who have popped in to add their expertise along the way. For me, the process has been exciting, and one that I hope will add value to looking for dolphin language.

And thank you to my editor, Joe Calamia at the University of Chicago Press, who not only patiently and carefully guided me through the changes in the publishing world since my last book but also meticulously edited my manuscript (more times than I want to remember) to make sure what I said was what I meant. Extra thanks to Matt Lang and others for help in the final phases of submission and with artwork.

As always, I owe my thanks to the dolphins, who, after forty years, still open their world to humans and allow us to observe their most intimate moments. Despite the impact of hurricanes and climate change on their habitat, the dolphins push to survive and pass on to their young what it means to be a healthy aquatic society.

Species Glossary

The following is a list of animals that appear in the book. Common names have been used in the book and are listed on the left. Scientific names are given on the right. Common names are listed in alphabetical order under each subheading.

Marine Mammals

Toothed Whales—Odontocetes—Dolphins and Porpoises

Atlantic spotted dolphin	*Stenella frontalis*
Atlantic white-sided dolphin	*Lagenorhynchus acutus*
beluga whale	*Delphinapterus leucas*
bottlenose dolphin	*Tursiops truncatus*
common dolphin	*Delphinus delphis*
Dall's porpoise	*Phocoenoides dalli*
false killer whale	*Pseudorca crassidens*
Guyana dolphin	*Sotalia guianensis*
harbor porpoise	*Phocoena phocoena*
Heaviside's dolphin	*Cephalorhynchus heavisidii*
Indian Ocean bottlenose dolphin	*Tursiops aduncus*
orca/killer whale	*Orcinus orca*

Pacific white-sided dolphin	*Lagenorhyncus obliquidens*
pantropical spotted dolphin	*Stenella attenuata*
pilot whale, long-finned	*Globicephala melas*
pilot whale, short-finned	*Globicephala macrorhynchus*
Risso's dolphin	*Grampus griseus*
sperm whale	*Physeter macrocephala*
spinner dolphin	*Stenella longirostris*
striped dolphin	*Stenella coeruleoalba*

Baleen Whales—Mysticetes

blue whale	*Balaenoptera musculus*
fin whale	*Balaenoptera physalus*
humpback whale	*Megaptera novaeangliae*
southern right whale	*Eubalaena australis*

Pinnipeds—Seals and Sea Lions

bearded seal	*Erignathus barbatus*
gray seal	*Halichoerus grypus*
harbor seal	*Phoca vitulina*
Mediterranean monk seal	*Monachus monachus*

Other Marine Mammals

otter	genus *Enhydra*
polar bear	*Ursus maritimus*

Primates—Apes and Monkeys

bonobo	*Pan paniscus*
Campbell's monkey	*Cercopithecus campbelli*
Common chimpanzee	*Pan troglodytes*
Diana monkey	*Cercopithecus diana*
hamadryas baboon	*Papio hamadryas*
macaque	genus *Macaca*
mountain gorilla	*Gorilla beringei graueri*
olive baboon	*Papio anubis*
orangutan	genus *Pongo*
vervet monkey	*Chlorocebus pygerythrus*
yellow-bellied marmoset	*Marmota flaviventris*

Birds

African gray parrot	*Psittacus erithacus*
American crows	family Corvidae
crowned hawk-eagle	*Stephanoaetus coronatus*
Darwin's finches	subfamily Geospizinae
domestic chick	*Gallus domesticus*
honeyguide birds	genus *Prodotiscus*
penguins	genus *Pinguinus*
raven	*Corvus corax*

Other Mammals

African elephant	*Loxodonta africana*
Asian elephant	*Elephas maximus*
bats	order Chiroptera
beavers	order Rodentia
coyote	*Canis latrans*
dingo	*Canis dingo*
dog	*Canis familiaris*
ground squirrel	*Spermophilus beecheyi*
hippo	*Hippopotamus amphibius*
horses	family Equidae
leopards	genus *Panthera*
lion	*Panthera leo*
mule deer	*Odocoileus hemionus*
oryx	genus *Oryx*
prairie dog	*Cynomys gunnisoni*
sika deer	*Cervus nippon*
zebras	genus *Equus*

Other Animals

African electric fish	genus *Electrophorus*
Aldabra tortoise	*Aldabrachelys gigantea*
bees	genus *Apis*
cephalopods	class Cephalopoda
manta rays	genus *Mobula*
shrimp	genus *Caridea*
turtles	family Cheloniidae

Notes

Preface

1. Goodall, Jane. *In the Shadow of Man*. London: Collins, 1971; Fossey, Dian. *Gorillas in the Mist*. Boston: Houghton Mifflin, 2000; Moss, Cynthia. *Elephant Memories: Thirteen Years in the Life of an Elephant Family*. New York: Morrow, 1988.

2. Musser, Whitney B., et al. "Differences in Acoustic Features of Vocalizations Produced by Killer Whales Cross-Socialized with Bottlenose Dolphins." *The Journal of the Acoustical Society of America* 136, no. 4 (2014): 1990–2002.

3. Seyfarth, Robert M., Dorothy L. Cheney, and Peter Marler. "Monkey Responses to Three Different Alarm Calls: Evidence of Predator Classification and Semantic Communication." *Science* 210, no. 4471 (1980): 801–803.

4. Slobodchikoff, Con. *Chasing Doctor Dolittle: Learning the Language of Animals*. New York: St. Martin's Press, 2012.

5. Munn, Charles A. "The Deceptive Use of Alarm Calls by Sentinel Species in Mixed Species Flocks of Neotropical Birds." In *Deception: Perspectives on Human and Nonhuman Deceit*, edited by Robert W. Mitchell and Nicholas S. Thompson, 169–176. Albany: State University of New York Press, 1986.

6. Ford, John Kenneth Baker. "Call Traditions and Dialects of Killer Whales (*Orcinus orca*) in British Columbia." PhD diss., University of British Columbia, 1984.

7. Mäthger, Lydia M., Nadav Shashar, and Roger T. Hanlon. "Do Cephalopods Communicate Using Polarized Light Reflections from Their Skin?" *Journal of Experimental Biology* 212, no. 14 (2009): 2133–2140.

8. Ireland, Hamish M., and Graeme D. Ruxton. "Zebra Stripes: An Interspecies Signal to Facilitate Mixed-Species Herding?" *Biological Journal of the Linnean Society* 121, no. 4 (2017): 947–952.

9. Kohlsdorf, Daniel, Denise Herzing, and Thad Starner. "Method for Discovering Models of Behavior: A Case Study with Wild Atlantic Spotted Dolphins." *Animal Behavior and Cognition* 3, no. 4 (2016): 265–287.

10. Whitehead, Hal, and Luke Rendell. *The Cultural Lives of Whales and Dolphins.* Chicago: University of Chicago Press, 2014.

11. Whorf, Benjamin Lee. *Language, Thought, and Reality: Selected Writings of Benjamin Lee Whorf.* Cambridge, MA: MIT Press, 2012.

12. Naess, Arne, and Satish Kumar. *Deep Ecology.* London: Phil Shepherd Production, 1992.

Chapter 1

1. Pryor, Karen, and Jon Lindbergh. "A Dolphin-Human Fishing Cooperative in Brazil." *Marine Mammal Science* 6, no. 1 (1990): 77–82.

2. Simões-Lopes, Paulo C., Marta E. Fabián, and João O. Menegheti. "Dolphin Interactions with the Mullet Artisanal Fishing on Southern Brazil: A Qualitative and Quantitative Approach." *Revista Brasileira de Zoologia* 15, no. 3 (1998): 709–726.

3. Isack, H. A., and H.-U. Reyer. "Honeyguides and Honey Gatherers: Interspecific Communication in a Symbiotic Relationship." *Science* 243, no. 4896 (1989): 1343–1346.

4. Fossey, Dian. *Gorillas in the Mist.* Boston: Houghton Mifflin, 2000, 14.

5. Ralls, Katherine, Patricia Fiorelli, and Sheri Gish. "Vocalizations and Vocal Mimicry in Captive Harbor Seals, *Phoca vitulina.*" *Canadian Journal of Zoology* 63, no. 5 (1985): 1050–1056.

6. Ridgway, Sam, et al. "Spontaneous Human Speech Mimicry by a Cetacean." *Current Biology* 22, no. 20 (2012): R860–R861.

7. Alves, Ana, et al. "Vocal Matching of Naval Sonar Signals by Long-Finned Pilot Whales (*Globicephala melas*)." *Marine Mammal Science* 30, no. 3 (2014): 1248–1257.

8. Stoeger, Angela S., et al. "An Asian Elephant Imitates Human Speech." *Current Biology* 22, no. 22 (2012): 2144–2148.

9. Wich, Serge A., et al. "A Case of Spontaneous Acquisition of a Human Sound by an Orangutan." *Primates* 50, no. 1 (2009): 56–64.

Chapter 2

1. Seyfarth, Robert M., Dorothy L. Cheney, and Peter Marler. "Monkey Responses to Three Different Alarm Calls: Evidence of Predator Classification and Semantic Communication." *Science* 210, no. 4471 (1980): 801–803.

2. Slobodchikoff, Constantine N., et al. "Semantic Information Distinguishing Individual Predators in the Alarm Calls of Gunnison's Prairie Dogs." *Animal Behaviour* 42, no. 5 (1991): 713–719.

3. de Waal, Frans. *Chimpanzee Politics: Power and Sex among Apes.* Baltimore: Johns Hopkins University Press, 2007.

4. Owings, Donald H., and Ross A. Virginia. "Alarm Calls of California Ground Squirrels (*Spermophilus beecheyi*)." *Zeitschrift für Tierpsychologie* 46, no. 1 (1978): 58–70.

5. Carrasco, Malle F., and Blumstein, Daniel T., 2012. "Mule Deer (*Odocoileus hemionus*) Respond to Yellow-Bellied Marmot (*Marmota flaviventris*) Alarm Calls." *Ethology* 118, no. 3 (2012): 243–250.

6. Griffin, Donald Redfield. *The Question of Animal Awareness: Evolutionary Continuity of Mental Experience*. New York: Rockefeller University Press, 1976.

7. Janik, Vincent M. "Cetacean Vocal Learning and Communication." *Current Opinion in Neurobiology* 28 (2014): 60–65.

8. Morton, Eugene S. "On the Occurrence and Significance of Motivation-Structural Rules in Some Bird and Mammal Sounds." *American Naturalist* 111, no. 981 (1977): 855–869.

9. Vakoch, Douglas A., and Jeffrey Punske, eds. *Xenolinguistics: Towards a Science of Extraterrestrial Language*. New York: Routledge, 2024.

10. Yong, Ed. *An Immense World: How Animal Senses Reveal the Hidden Realms Around Us*. Toronto: Knopf Canada, 2022.

11. Von Frisch, Karl. "Dialects in the Language of the Bees." *Scientific American* 207, no. 2 (1962): 78–89; Rosin, Ruth. "The Honey-Bee 'Dance Language' Hypothesis and the Foundations of Biology and Behavior." *Journal of Theoretical Biology* 87, no. 3 (1980): 457–481.

12. Hanlon, Roger. "Cephalopod Dynamic Camouflage." *Current Biology* 17, no. 11 (2007): R400–R404.

13. Rößler, Helen, et al. "Are Icelandic Harbor Seals Acoustically Cryptic to Avoid Predation?" *JASA Express Letters* 1, no. 3 (2021).

14. Barrett-Lennard, Lance G., John K. B. Ford, and Kathy A. Heise. "The Mixed Blessing of Echolocation: Differences in Sonar Use by Fish-Eating and Mammal-Eating Killer Whales." *Animal Behaviour* 51, no. 3 (1996): 553–565.

15. Martin, Morgan J., et al. "Heaviside's Dolphins (*Cephalorhynchus heavisidii*) Relax Acoustic Crypsis to Increase Communication Range." *Proceedings of the Royal Society B: Biological Sciences* 285, no. 1883 (2018): 20181178.

16. Payne, Katy. "Eavesdropping on Elephants." *Journal of the Acoustical Society of America* 115, no. 5 supplement (2004): 2553–2554; Mott, Maryann. "Did Animals Sense Tsunami Was Coming?" *National Geographic*, January 4, 2005. https://www.nationalgeographic.com/animals/article/news-animals-tsunami-sense-coming.

17. Filatova, O. A., A. M. Burdin, and E. Hoyt. "Horizontal Transmission of Vocal Traditions in Killer Whale (*Orcinus orca*) Dialects." *Biology Bulletin* 37, no. 9 (2010): 965–971.

18. Bender, Courtney E., Denise L. Herzing, and David F. Bjorklund. "Evidence of Teaching in Atlantic Spotted Dolphins (*Stenella frontalis*) by Mother Dolphins Foraging in the Presence of Their Calves." *Animal Cognition* 12, no. 1 (2009): 43–53.

19. De Brabanter, Gaïane L. B., Denise L. Herzing, and Susan Jarvis. "Exploration of Horizontal Information Transmission through Social Learning in Juvenile Atlantic Spotted Dolphins (*Stenella frontalis*)." *Animal Behavior and Cognition* 4, no. 4 (2017): 425–441.

20. Wells, Randall. S. "Social Structure and Life History of Bottlenose Dolphins near Sarasota Bay, Florida: Insights from Four Decades and Five Generations." In *Primates and Cetaceans: Field Research and Conservation of Complex Mammalian Societies*, edited by J. Yamagiwa and L. Karczmarski, 149–172. Tokyo: Springer, 2014.

21. Slobodchikoff, C. N., Bianca S. Perla, and Jennifer L. Verdolin. *Prairie Dogs: Communication and Community in an Animal Society.* Cambridge, MA: Harvard University Press, 2009.

22. Caldwell, Melba C., David K. Caldwell, and Peter L. Tyack. "Review of the Signature-Whistle Hypothesis for the Atlantic Bottlenose Dolphin." In *The Bottlenose Dolphin,* edited by Stephen Leatherwood and Randall R. Reeves, 199–234. Academic Press, 1990.

23. Kershenbaum, Arik, Laela S. Sayigh, and Vincent M. Janik. "The Encoding of Individual Identity in Dolphin Signature Whistles: How Much Information Is Needed?" *PloS One* 8, no. 10 (2013): e77671.

24. Quick, Nicola J., and Vincent M. Janik. "Bottlenose Dolphins Exchange Signature Whistles When Meeting at Sea." *Proceedings of the Royal Society B: Biological Sciences* 279, no. 1738 (2012): 2539–2545.

25. King, Stephanie L., and Vincent M. Janik. "Bottlenose Dolphins Can Use Learned Vocal Labels to Address Each Other." *Proceedings of the National Academy of Sciences* 110, no. 32 (2013): 13216–13221.

26. Lilly, John C., and Alice M. Miller. "Vocal Exchanges between Dolphins: Bottlenose Dolphins 'Talk' to Each Other with Whistles, Clicks, and a Variety of Other Noises." *Science* 134, no. 3493 (1961): 1873–1876; Bastian, Jarvis. *The Transmission of Arbitrary Environmental Information between Bottle-Nose Dolphins.* NOTS Technical Publication 4117. China Lake, CA: US Naval Ordnance Test Station, 1967.

27. Eskelinen, Holli C., et al. "Acoustic Behavior Associated with Cooperative Task Success in Bottlenose Dolphins (*Tursiops truncatus*)." *Animal Cognition* 19, no. 4 (2016): 789–797.

28. Kabadayi, Can, and Mathias Osvath. "Ravens Parallel Great Apes in Flexible Planning for Tool-Use and Bartering." *Science* 357, no. 6347 (2017): 202–204; Pika, Simone, and Thomas Bugnyar. "The Use of Referential Gestures in Ravens (*Corvus corax*) in the Wild." *Nature Communications* 2, no. 1 (2011): 1–5.

29. Baron, Susan C., et al. "Differences in Acoustic Signals from Delphinids in the Western North Atlantic and Northern Gulf of Mexico." *Marine Mammal Science* 24, no. 1 (2008): 42–56.

30. Bazúa-Durán, Carmen, Whitlow W. L. Au, and Julie N. Oswald. "Geographic Variations in the Whistles of Spinner Dolphins (*Stenella longirostris*)." *Journal of the Acoustical Society of America* 112, no. 5_Supplement (2002): 2400-2400.

31. Ford, John K. B., and H. Dean Fisher. "Group-Specific Dialects of Killer Whales (*Orcinus orca*) in British Columbia." *Communication and Behavior of Whales* 76 (1983): 129.

32. May-Collado, Laura J. "Changes in Whistle Structure of Two Dolphin Species during Interspecific Associations." *Ethology* 116, no. 11 (2010): 1065–1074.

33. Musser, Whitney B., et al. "Differences in Acoustic Features of Vocalizations Produced by Killer Whales Cross-Socialized with Bottlenose Dolphins." *The Journal of the Acoustical Society of America* 136, no. 4 (2014): 1990–2002; Cosentino, Mel, et al. "I Beg Your Pardon? Acoustic Behaviour of a Wild Solitary Common Dolphin Who Interacts with Harbour Porpoises." *Bioacoustics* 31, no. 5 (2022): 517–534.

34. Déaux, Éloïse C., Isabelle Charrier, and Jennifer A. Clarke. "The Bark, the Howl and the Bark-Howl: Identity Cues in Dingoes' Multicomponent Calls." *Behavioural Processes* 129 (2016): 94–100; Charrier, Isabelle, et al. "Individual Signatures in the Vocal Repertoire of the Endangered Mediterranean Monk Seal: New Perspectives for Population Monitoring." *Endangered Species Research* 32 (2017): 459–470; Ford, John Kenneth Baker. "Call Traditions and Dialects of Killer Whales (*Orcinus orca*) in British Columbia." PhD diss., University of British Columbia, 1984.

35. King, Barbara J. *The Dynamic Dance: Nonvocal Communication in African Great Apes*. Cambridge, MA: Harvard University Press, 2009.

36. Arcadi, Adam Clark. "Phrase Structure of Wild Chimpanzee Pant Hoots: Patterns of Production and Interpopulation Variability." *American Journal of Primatology* 39, no. 3 (1996): 159–178.

37. Seyfarth, Robert. M., and Dorothy L. Cheney. "Vocal Development in Vervet Monkeys." *Animal Behaviour* 34, no. 6 (1986): 1640–1658.

38. Chomsky, Noam. "The General Properties of Language." In *Brain Mechanisms Underlying Speech and Language*, edited by F. L. Darley, 73–88. New York: Grune and Stratton, 1967.

39. Janik, Vincent M. "Cetacean Vocal Learning and Communication." *Current Opinion in Neurobiology* 28 (2014): 60–65.

40. Deecke, Volker B., John K. B. Ford, and Paul Spong. "Quantifying Complex Patterns of Bioacoustic Variation: Use of a Neural Network to Compare Killer Whale (*Orcinus orca*) Dialects." *The Journal of the Acoustical Society of America* 105, no. 4 (1999): 2499–2507.

41. Kohlsdorf, Daniel, et al. "Probabilistic Extraction and Discovery of Fundamental Units in Dolphin Whistles." *2014 IEEE International Conference on Acoustics, Speech and Signal Processing (ICASSP)*, 8242–8246. Florence: IEEE, 2014.

42. Kohlsdorf, Daniel, et al. "An Underwater Wearable Computer for Two-Way Human-Dolphin Communication Experimentation." *Proceedings of the 2013 International Symposium on Wearable Computers*, 147–148. New York: Association for Computing Machinery, 2013.

43. Herzing, Denise L. *Dolphin Diaries: My 25 Years with Spotted Dolphins in the Bahamas*. New York: St. Martin's Griffin, 2011.

44. Herzing, Denise L., et al. "Exodus! Large-Scale Displacement and Social Adjustments of Resident Atlantic Spotted Dolphins (*Stenella frontalis*) in the Bahamas." *PloS One* 12, no. 8 (2017): e0180304.

Chapter 3

1. Herzing, Denise L., and Christine M. Johnson. "Interspecific Interactions between Atlantic Spotted Dolphins (*Stenella frontalis*) and Bottlenose Dolphins (*Tursiops truncatus*) in the Bahamas." *Aquatic Mammals* 23 (1997): 85–99.

2. Xitco, Mark J., and Herbert L. Roitblat. "Object Recognition through Eavesdropping: Passive Echolocation in Bottlenose Dolphins." *Animal Learning & Behavior* 24, no. 4 (1996): 355–365.

3. Magrath, Robert D., et al. "Eavesdropping on Heterospecific Alarm Calls: From Mechanisms to Consequences." *Biological Reviews* 90, no. 2 (2015): 560–586.

4. Volker, Cassie, and Denise Herzing. "Aggressive Behaviors of Adult Male Atlantic Spotted Dolphins: Making Signals Count during Intraspecific and Interspecific Conflicts." *Animal Behavior and Cognition* 8, no. 1 (2021): 36–51.

5. Smith, W. John. "Signaling Behavior: Contributions of Different Repertoires." In *Dolphin Cognition and Behavior: A Comparative Approach*, edited by Ronald J. Schusterman, Jeanette A. Thomas, and Forrest G. Wood, 315. New York: Psychology Press, 1986.

6. Morton, Eugene S. "On the Occurrence and Significance of Motivation-Structural Rules in Some Bird and Mammal Sounds." *American Naturalist* 111, no. 981 (1977): 855–869.

7. Holland, Jennifer S. *Unlikely Friendships: 47 Remarkable Stories from the Animal Kingdom*. New York: Workman Publishing, 2011.

8. Fouts, Roger S., and Joseph B. Couch. "Cultural Evolution of Learned Language in Chimpanzees." In *Communicative Behavior and Evolution*, edited by Martin E. Hahn and Edward C. Simmel, 141–161. New York: Academic Press, 1976; Gardner, R. Allen, and Beatrice T. Gardner. "Teaching Sign Language to a Chimpanzee." *Science* 165, no. 3894 (1969): 664–672; Miles, Lyn W. "Language Acquisition in Apes and Children." In *Sign Language and Language Acquisition in Man and Ape*, edited by Fred C. C. Peng, 103–120. London: Routledge, 2019; Patterson, Francine G. "The Gestures of a Gorilla: Language Acquisition in Another Pongid." *Brain and Language* 5, no. 1 (1978): 72–97.

9. Herman, Louis M., Adam A. Pack, and Palmer Morrel-Samuels. "Representational and Conceptual Skills of Dolphins." In *Language and Communication: Comparative Perspectives*, edited by Herbert L. Roitblat, Louis M. Herman, and Paul E. Nachtigall, 403–442. New York: Psychology Press, 1993; Marten, Ken, and Suchi Psarakos. "Using Self-View Television to Distinguish between Self-Examination and Social Behavior in the Bottlenose Dolphin (*Tursiops truncatus*)." *Consciousness and Cognition* 4, no. 2 (1995): 205–224; Reiss, Diana, and Brenda McCowan. "Spontaneous Vocal Mimicry and Production by Bottlenose Dolphins (*Tursiops truncatus*): Evidence for Vocal Learning." *Journal of Comparative Psychology* 107, no. 3 (1993): 301.

10. Pepperberg, Irene M. "Acquisition of Anomalous Communicatory Systems: Implications for Studies on Interspecies Communication." *Dolphin Cognition and Behavior: A Comparative Approach*, edited by Ronald J. Schusterman, Jeanette A. Thomas, and Forrest G. Wood, 289–302. New York: Psychology Press, 1986.

11. Magrath, Robert D., et al. "Eavesdropping on Heterospecific Alarm Calls: From Mechanisms to Consequences." *Biological Reviews* 90, no. 2 (2015): 560–586.

12. Koda, Hiroki. "Possible Use of Heterospecific Food-Associated Calls of Macaques by Sika Deer for Foraging Efficiency." *Behavioural Processes* 91, no. 1 (2012): 30–34.

13. Zuberbühler, Klaus. "Interspecies Semantic Communication in Two Forest Primates." *Proceedings of the Royal Society of London. Series B: Biological Sciences* 267, no. 1444 (2000): 713–718.

14. Igic, Branislav, et al. "Crying Wolf to a Predator: Deceptive Vocal Mimicry by a Bird Protecting Young." *Proceedings of the Royal Society B: Biological*

Sciences 282, no. 1809 (2015): 20150798; Ghoul, Melanie, Ashleigh S. Griffin, and Stuart A. West. "Toward an Evolutionary Definition of Cheating." *Evolution* 68 no. 2 (2014): 318–331.

15. Kimura, Yukiko. "War Brides in Hawai'i and Their In-Laws." *American Journal of Sociology* 63, no. 1 (1957): 70–76.

16. Blanke, Detlev. "Causes of the Relative Success of Esperanto." *Language Problems and Language Planning* 33, no. 3 (2009): 251–266.

17. May-Collado, Laura J. "Changes in Whistle Structure of Two Dolphin Species during Interspecific Associations." *Ethology* 116, no. 11 (2010): 1065–1074; Cosentino, Mel, et al. "I Beg Your Pardon? Acoustic Behaviour of a Wild Solitary Common Dolphin Who Interacts with Harbour Porpoises." *Bioacoustics* 31, no. 5 (2022): 517–534.

18. Ford, John K. B., and H. Dean Fisher. "Group-Specific Dialects of Killer Whales (*Orcinus orca*) in British Columbia." *Communication and Behavior of Whales* 76 (1983): 129.

19. Herman, Louis M., Adam A. Pack, and Palmer Morrel-Samuels. "Representational and Conceptual Skills of Dolphins." In *Language and Communication: Comparative Perspectives*, edited by Herbert L. Roitblat, Louis M. Herman, and Paul E. Nachtigall, 403–442. New York: Psychology Press, 1993.

20. Xitco, Mark J. "Mimicry of Modeled Behaviors by Bottlenose Dolphins." Master Thesis, University of Hawai'i at Manoa, 1988.

21. Hopkin, Michael. "Elephants Do Impressions." *Nature*, March 23, 2005.

22. Ridgway, Sam, et al. "Spontaneous Human Speech Mimicry by a Cetacean." *Current Biology* 22, no. 20 (2012): R860–R861.

23. Richards, Douglas G., James P. Wolz, and Louis M. Herman. "Vocal Mimicry of Computer-Generated Sounds and Vocal Labeling of Objects by a Bottlenosed Dolphin, *Tursiops Truncatus*." *Journal of Comparative Psychology* 98, no. 1 (1984): 10.

24. Panova, Elena M., and Alexandr V. Agafonov. "A Beluga Whale Socialized with Bottlenose Dolphins Imitates Their Whistles." *Animal Cognition* 20, no. 6 (2017): 1153–1160.

25. Musser, Whitney B., et al. "Differences in Acoustic Features of Vocalizations Produced by Killer Whales Cross-Socialized with Bottlenose Dolphins." *The Journal of the Acoustical Society of America* 136, no. 4 (2014): 1990–2002.

26. Ridgway, Sam, et al. "Spontaneous Human Speech Mimicry by a Cetacean." *Current Biology* 22, no. 20 (2012): R860–R861.

27. Abramson, José Z., et al. "Imitation of Novel Conspecific and Human Speech Sounds in the Killer Whale (*Orcinus orca*)." *Proceedings of the Royal Society B: Biological Sciences* 285, no. 1871 (2018): 20172171.

28. Connor, Richard C., Rachel Smolker, and Lars Bejder. "Synchrony, Social Behaviour and Alliance Affiliation in Indian Ocean Bottlenose Dolphins, *Tursiops aduncus*." *Animal Behaviour* 72, no. 6 (2006): 1371–1378.

29. Myers, Alyson J., Denise L. Herzing, and David F. Bjorklund. "Synchrony during Aggression in Adult Male Atlantic Spotted Dolphins (*Stenella frontalis*)." *Acta Ethologica* 20, no. 2 (2017): 175–185.

30. Tyack, Peter L. "Convergence of Calls as Animals Form Social Bonds, Active Compensation for Noisy Communication Channels, and the Evolution

of Vocal Learning in Mammals." *Journal of Comparative Psychology* 122, no. 3 (2008): 319.

31. Thaut, Michael H. "Entrainment and the Motor System." *Music Therapy Perspectives* 31, no. 1 (2013): 31–34.

32. Wilson, Margaret, and Peter F. Cook. "Rhythmic Entrainment: Why Humans Want to, Fireflies Can't Help It, Pet Birds Try, and Sea Lions Have to Be Bribed." *Psychonomic Bulletin & Review* 23, no. 6 (2016): 1647–1659.

33. Balconi, Michela, and Maria Elide Vanutelli. "Vocal and Visual Stimulation, Congruence and Lateralization Affect Brain Oscillations in Interspecies Emotional Positive and Negative Interactions." *Social Neuroscience* 11, no. 3 (2016): 297–310.

34. Kinreich, Sivan, et al. "Brain-to-Brain Synchrony during Naturalistic Social Interactions." *Scientific Reports* 7, no. 1 (2017): 1–12.

35. Isack, H. A., and H.-U. Reyer. "Honeyguides and Honey Gatherers: Interspecific Communication in a Symbiotic Relationship." *Science* 243, no. 4896 (1989): 1343–1346.

36. Spottiswoode, Claire N., and Brian M. Wood. "Culturally Determined Interspecies Communication between Humans and Honeyguides." *Science* 382, no. 6675 (2023): 1155–1158.

37. Pryor, Karen, and Jon Lindbergh. "A Dolphin-Human Fishing Cooperative in Brazil." *Marine Mammal Science* 6, no. 1 (1990): 77–82.

38. Hatkoff, Craig, and Isabella Hatkoff. *Owen and Mzee: The True Story of a Remarkable Friendship*. New York: Scholastic Inc., 2006.

39. Munene, Mugumo. "Surprise in the Kenyan Wild as Lioness Adopts Oryx." *Daily Nation News*, January 7, 2002.

40. Bérubé, Martine, and Alex Aguilar. "A New Hybrid between a Blue Whale, *Balaenoptera musculus*, and a Fin Whale, *B. physalus*: Frequency and Implications of Hybridization." *Marine Mammal Science* 14, no. 1 (1998): 82–98.

41. Sylvestre, Jean-Pierre, and Soichi Tasaka. "On the Intergeneric Hybrids in Cetaceans." *Aquatic Mammals* 11, no. 3 (1985): 101–108.

42. Psarakos, Suchi, Denise L. Herzing, and Ken Marten. "Mixed-Species Associations between Pantropical Spotted Dolphins (*Stenella attenuata*) and Hawai'ian Spinner Dolphins (*Stenella longirostris*) off Oahu, Hawai'i." *Aquatic Mammals* 29, no. 3 (2003): 390–395.

43. Strier, Karen B. *Primate Behavioral Ecology*. New York: Routledge, 2015.

44. Wolters, Sonja, and Klaus Zuberbühler. "Mixed-Species Associations of Diana and Campbell's Monkeys: The Costs and Benefits of a Forest Phenomenon." *Behaviour* 140, no. 3 (2003): 371–385.

45. Willis, Pamela M., et al. "Natural Hybridization between Dall's Porpoises (*Phocoenoides dalli*) and Harbour Porpoises (*Phocoena phocoena*)." *Canadian Journal of Zoology* 82, no. 5 (2004): 828–834.

46. Herzing, Denise L., Kelly Moewe, and Barbara J. Brunnick. "Interspecies Interactions between Atlantic Spotted Dolphins, *Stenella frontalis* and Bottlenose Dolphins, *Tursiops truncatus*." *Aquatic mammals* 29 (2003): 335–341.

47. Palmer, Daniela H., and Marcus R. Kronforst. "Divergence and Gene Flow among Darwin's Finches: A Genome-Wide View of Adaptive Radiation Driven by Interspecies Allele Sharing." *Bioessays* 37, no. 9 (2015): 968–974.

48. Frantzis, Alexandros, and Denise L. Herzing. "Mixed-Species Associations of Striped Dolphins (*Stenella coeruleoalba*), Short-Beaked Common Dolphins (*Delphinus delphis*), and Risso's Dolphins (*Grampus griseus*) in the Gulf of Corinth (Greece, Mediterranean Sea)." *Aquatic Mammals* 28, no. 2 (2002): 188–197.

49. Tonay, Arda M., Ayhan Dede, and Ayaka Amaha Öztürk. "An Unusual Interaction between Bottlenose Dolphins (*Tursiops truncatus*) and a Harbour Porpoise (*Phocoena phocoena*)." *J. Black Sea/Mediterranean Environment* 23, no. 3 (2017): 222–228.

50. Sakai, Mai, et al. "A Wild Indo-Pacific Bottlenose Dolphin Adopts a Socially and Genetically Distant Neonate." *Scientific Reports* 6, no. 1 (2016): 1–8.

51. Carzon, Pamela, et al. "Cross-Genus Adoptions in Delphinids: One Example with Taxonomic Discussion." *Ethology* 125, no. 9 (2019): 669–676.

52. Brasseur, I., et al. "New Results on a Bottlenose Dolphin (*Tursiops truncatus*) Community at Rangiroa Island (French Polynesia)." *The 16th Annual Conference of the European Cetacean Society*, Liège, Belgium, April 8–10, 2002.

53. Mrusczok, Marie-Thérèse, et al. "First Account of Apparent Alloparental Care of a Long-Finned Pilot Whale Calf (*Globicephala melas*) by a Female Killer Whale (*Orcinus orca*)." *Canadian Journal of Zoology* 101, no. 4 (2023): 288–293.

54. Syme, Jonathan, Jeremy J. Kiszka, and Guido J. Parra. "Multiple Social Benefits Drive the Formation of Mixed-Species Groups of Australian Humpback and Indo-Pacific Bottlenose Dolphins." *Behavioral Ecology and Sociobiology* 77, no. 4 (2023): 43.

55. Pitman, Robert L., et al. "Humpback Whales Interfering When Mammal-Eating Killer Whales Attack Other Species: Mobbing Behavior and Interspecific Altruism?" *Marine Mammal Science* 33, no. 1 (2017): 7–58.

56. Connor, Richard C., and Kenneth S. Norris. "Are Dolphins Reciprocal Altruists?" *American Naturalist* 119, no. 3 (1982): 358–374; Whitehead, Hal, and Luke Rendell. *The Cultural Lives of Whales and Dolphins*. Chicago: University of Chicago Press, 2014.

57. Pate, Jessica H., and Andrea D. Marshall. "Urban Manta Rays: Potential Manta Ray Nursery Habitat along a Highly Developed Florida Coastline." *Endangered Species Research* 43 (2020): 51–64.

58. Fimrite, Peter. "Daring Rescue of Whale off Farallones/Humpback Nuzzled Her Saviors in Thanks After They Untangled Her from Crab Lines, Diver Says." *San Francisco Gate*, December 14, 2005.

59. Pérez-Manrique, Ana, and Antoni Gomila. "The Comparative Study of Empathy: Sympathetic Concern and Empathic Perspective-Taking in Non-human Animals." *Biological Reviews* 93, no. 1 (2018): 248–269.

60. Madsen, Elainie Alenkær, and Tomas Persson. "Contagious Yawning in Domestic Dog Puppies (*Canis lupus familiaris*): The Effect of Ontogeny and Emotional Closeness on Low-Level Imitation in Dogs." *Animal Cognition* 16, no. 2 (2013): 233–240; Avarguès-Weber, Aurore, Erika H. Dawson, and Lars Chittka. "Mechanisms of Social Learning across Species Boundaries." *Journal of Zoology* 290, no. 1 (2013): 1–11.

61. Stamenov, Maksim, and Vittorio Gallese, eds. *Mirror Neurons and the Evolution of Brain and Language*. Vol. 42. Amsterdam: John Benjamins Publishing, 2002.

Chapter 4

1. Davis, Hank Ed, and Dianne A. Balfour. *The Inevitable Bond: Examining Scientist-Animal Interactions*. Cambridge: Cambridge University Press, 1992.

2. Miklósi, Ádám, József Topál, and Vilmos Csányi. "Comparative Social Cognition: What Can Dogs Teach Us?" *Animal Behaviour* 67, no. 6 (2004): 995–1004.

3. Pongrácz, Péter, et al. "Human Listeners Are Able to Classify Dog (*Canis familiaris*) Barks Recorded in Different Situations." *Journal of Comparative Psychology* 119, no. 2 (2005): 136.

4. Jackson, Melody M., et al. "FIDO—Facilitating Interactions for Dogs with Occupations: Wearable Communication Interfaces for Working Dogs." *Personal and Ubiquitous Computing* 19, no. 1 (2015): 155–173.

5. Pepperberg, Irene M. "Cognitive and Communicative Abilities of Grey Parrots." *Applied Animal Behaviour Science* 100, nos. 1–2 (2006): 77–86.

6. Savage-Rumbaugh, E. Sue, and Roger Lewin. *Kanzi: The Ape at the Brink of the Human Mind*. New York: Wiley, 1994.

7. Pedersen, Janni, and William M. Fields. "Aspects of Repetition in Bonobo–Human Conversation: Creating Cohesion in a Conversation between Species." *Integrative Psychological and Behavioral Science* 43, no. 1 (2009): 22–41.

8. Lorenz, Konrad. *King Solomon's Ring: New Light on Animal Ways*. London: Psychology Press, 2002.

9. Hayes, Keith J., and Catherine Hayes. "Imitation in a Home-Raised Chimpanzee." *Journal of Comparative and Physiological Psychology* 45, no. 5 (1952): 450.

10. Premack, David. "Language and Intelligence in Ape and Man: How Much Is the Gap between Human and Animal Intelligence Narrowed by Recent Demonstrations of Language in Chimpanzees?" *American Scientist* 64, no. 6 (1976): 674–683.

11. Terrace, Herbert S., et al. "Can an Ape Create a Sentence?" *Science* 206, no. 4421 (1979): 891–902.

12. Gardner, R. Allen, and Beatrice T. Gardner. "Teaching Sign Language to a Chimpanzee." *Science* 165, no. 3894 (1969): 664–672.

13. Pfungst, Oskar. *Clever Hans (the Horse of Mr. Von Osten): A Contribution to Experimental Animal and Human Psychology*. New York: Holt, Rinehart & Winston, 1911; Prescott, John H. "Clever Hans: Training the Trainers, or the Potential for Misinterpreting the Results of Dolphin Research." *NYASA* 364, no. 1 (1981): 130–136.

14. Fouts, Roger S., and Joseph B. Couch. "Cultural Evolution of Learned Language in Chimpanzees." In *Communicative Behavior and Evolution*, edited by Martin E. Hahn and Edward C. Simmel, 141–161. New York: Academic Press, 1976.

15. Miles, H. Lyn. "The Cognitive Foundations for Reference in a Signing Orangutan." In *Language and Intelligence in Monkeys and Apes: Comparative Developmental Perspectives*, edited by S. T. Parker and K. R. Gibson, 511–539. Cambridge: Cambridge University Press, 1990.

16. Rumbaugh, Duane M., and Timothy V. Gill. "Lana's Acquisition of Language Skills." In *Language Learning by a Chimpanzee*, edited by Duane M. Rumbaugh, 165–192. New York: Academic Press, 1977.

17. Savage-Rumbaugh, E. Sue, Duane M. Rumbaugh, and Sally Boysen. "Linguistically Mediated Tool Use and Exchange by Chimpanzees (*Pan troglodytes*)." In *Speaking of Apes*, edited by Thomas A. Sebeok and Jean Umiker-Sebeok, 353–383. Boston: Springer, 1980.

18. Savage-Rumbaugh, E. Sue, and Roger Lewin. *Kanzi: The Ape at the Brink of the Human Mind.* New York: Wiley, 1994.

19. Patterson, Francine G. "The Gestures of a Gorilla: Language Acquisition in Another Pongid." *Brain and Language* 5, no. 1 (1978): 72–97.

20. Herman, Louis M. "Cognition and Language Competencies of Bottle-nosed Dolphins." In *Dolphin Cognition and Behavior: A Comparative Approach*, edited by Ronald J. Schusterman, Jeanette A. Thomas, and Forrest G. Wood, 221–252. New York: Psychology Press, 1986.

21. Batteau, Dwight W. *Man/Dolphin Communication: Final Report 15 December 1966–13 December 1967.* Arlington, MA: LISTENING, Incorporated, 1967.

22. Lilly, John C. *Communication between Man and Dolphin.* New York: Crown Publishers, 1978.

23. Herman, Louis M., Adam A. Pack, and Palmer Morrel-Samuels. "Representational and Conceptual Skills of Dolphins." In *Language and Communication: Comparative Perspectives*, edited by Herbert L. Roitblat, Louis M. Herman, and Paul E. Nachtigall, 403–442. New York: Psychology Press, 1993.

24. Reiss, Diana, and Brenda McCowan. "Spontaneous Vocal Mimicry and Production by Bottlenose Dolphins (*Tursiops truncatus*): Evidence for Vocal Learning." *Journal of Comparative Psychology* 107, no. 3 (1993): 301.

25. Delfour, Fabienne, and Ken Marten. "Inter-Modal Learning Task in Bottlenosed Dolphins (*Tursiops truncatus*): A Preliminary Study Showed That Social Factors Might Influence Learning Strategies." *Acta Ethologica* 8, no. 1 (2005): 57–64.

26. Xitco, Mark J., John D. Gory, and Stan A. Kuczaj. "Spontaneous Pointing by Bottlenose Dolphins (*Tursiops truncatus*)." *Animal Cognition* 4, no. 2 (2001): 115–123.

27. Herzing, Denise L., Fabienne Delfour, and Adam A. Pack. "Responses of Human-Habituated Wild Atlantic Spotted Dolphins to Play Behaviors Using a Two-Way Human/Dolphin Interface." *International Journal of Comparative Psychology* 25, no. 2 (2012).

28. Turner, Jack. *The Abstract Wild.* Tucson: University of Arizona Press, 1996.

29. Amundin, Mats, et al. "An Echolocation Visualization and Interface System for Dolphin Research." *The Journal of the Acoustical Society of America* 123, no. 2 (2008): 1188–1194.

30. Kohlsdorf, Daniel, et al. "An Underwater Wearable Computer for Two Way Human-Dolphin Communication Experimentation." In *Proceedings of the 2013 International Symposium on Wearable Computers*, 147–148. New York: Association for Computing Machinery, 2013.

31. Kohlsdorf, Daniel, Denise Herzing, and Thad Starner. "Method for Discovering Models of Behavior: A Case Study with Wild Atlantic Spotted Dolphins." *Animal Behavior and Cognition* 3, no. 4 (2016): 265–287.

Chapter 5

1. Savage-Rumbaugh, E. Sue, and Roger Lewin. *Kanzi: The Ape at the Brink of the Human Mind*. New York: Wiley, 1994.

2. Slobodchikoff, Constantine N. *Chasing Doctor Dolittle: Learning the Language of Animals*. New York: St. Martin's Press, 2012.

3. Kohlsdorf, Daniel, et al. "Probabilistic Extraction and Discovery of Fundamental Units in Dolphin Whistles." *2014 IEEE International Conference on Acoustics, Speech and Signal Processing (ICASSP)*, 8242–8246. Florence: IEEE, 2014.

4. Sayigh, Laela, et al. "Repeated Call Types in Short-Finned Pilot Whales, *Globicephala macrorhynchus*." *Marine Mammal Science* 29, no. 2 (2013): 312–324.

5. Morton, Eugene S. "On the Occurrence and Significance of Motivation-Structural Rules in Some Bird and Mammal Sounds." *American Naturalist* 111, no. 981 (1977): 855–869.

6. Pongrácz, Péter, Csaba Molnár, and Ádám Miklósi. "Acoustic Parameters of Dog Barks Carry Emotional Information for Humans." *Applied Animal Behaviour Science* 100, no. 3–4 (2006): 228–240.

7. Koshiba, Mamiko, et al. "Familiarity Perception Call Elicited under Restricted Sensory Cues in Peer-Social Interactions of the Domestic Chick." *PloS one* 8, no. 3 (2013): e58847; Russo, Danilo, et al. "The Buzz of Drinking on the Wing in Echolocating Bats." *Ethology* 122, no. 3 (2016): 226–235.

8. Charrier, Isabelle, et al. "Individual Signatures in the Vocal Repertoire of the Endangered Mediterranean Monk Seal: New Perspectives for Population Monitoring." *Endangered Species Research* 32 (2017): 459–470.

9. Tinbergen, Niko. *The Animal in Its World (Explorations of an Ethologist, 1932–1972): Field Studies*. Vol. 84. Cambridge, MA: Harvard University Press, 1972; Lorenz, Konrad Z. *King Solomon's Ring: New Light on Animal Ways*. New York: Time Inc., 1962.

10. Volker, Cassie, and Denise Herzing. "Aggressive Behaviors of Adult Male Atlantic Spotted Dolphins: Making Signals Count during Intraspecific and Interspecific Conflicts." *Animal Behavior and Cognition* 8, no. 1 (2021): 36–51.

11. Hoffmann-Kuhnt, M., et al. "Whose Line Sound Is It Anyway? Identifying the Vocalizer on Underwater Video by Localizing with a Hydrophone Array." *Animal Behavior and Cognition* 3, no. 4 (2016): 288–298.

12. Mahoney, John R., et al. "How Hidden Are Hidden Processes? A Primer on Crypticity and Entropy Convergence." *Chaos: An Interdisciplinary Journal of Nonlinear Science* 21, no. 3 (2011).

13. Jones, Gareth. "Acoustic Signals and Speciation: The Roles of Natural and Sexual Selection in the Evolution of Cryptic Species." *Advances in the Study of Behaviour* 26 (1997): 317–354.

14. Bernard, H. Russell, and Clarence C. Gravlee, eds. *Handbook of Methods in Cultural Anthropology*. Lanham, MD: Rowman & Littlefield, 2014.

15. Favareau, Donald. "Life in the Co-operative Transformation Zone." *Tartu Semiotics Library* 19 (2018): 113–124.

16. Seyfarth, Robert M., Dorothy L. Cheney, and Peter Marler. "Monkey Responses to Three Different Alarm Calls: Evidence of Predator Classification and Semantic Communication." *Science* 210, no. 4471 (1980): 801–803.

17. Slobodchikoff, Constantine Nicholas, Bianca S. Perla, and Jennifer L. Verdolin. *Prairie Dogs: Communication and Community in an Animal Society.* Cambridge, MA: Harvard University Press, 2009.

18. Quick, Nicola J., and Vincent M. Janik. "Bottlenose Dolphins Exchange Signature Whistles When Meeting at Sea." *Proceedings of the Royal Society B: Biological Sciences* 279, no. 1738 (2012): 2539–2545.

Chapter 6

1. Deecke, Volker B., John K. B. Ford, and Paul Spong. "Quantifying Complex Patterns of Bioacoustic Variation: Use of a Neural Network to Compare Killer Whale (*Orcinus orca*) Dialects." *The Journal of the Acoustical Society of America* 105, no. 4 (1999): 2499–2507.

2. Sorensen, Scott, et al. "Deep Learning for Polar Bear Detection." In Image Analysis: 20th *Scandinavian Conference, SCIA 2017*, edited by Puneet Sharma and Filippo Maria Bianchi, 457–467. Cham, Switzerland: Springer, 2017.

3. Wiener, Seth, and Rory Turnbull. "Constraints of Tones, Vowels and Consonants on Lexical Selection in Mandarin Chinese." *Language and Speech* 59, no. 1 (2016): 59–82.

4. Meyer, Julien. "Typology and Acoustic Strategies of Whistled Languages: Phonetic Comparison and Perceptual Cues of Whistled Vowels." *Journal of the International Phonetic Association* 38, no. 1 (2008): 69–94.

5. Chiang, Wen-yu, and Fang-mei Chiang. "Saisiyat as a Pitch Accent Language: Evidence from Acoustic Study of Words." *Oceanic Linguistics* 44, no. 2 (2005): 404–426.

6. Cummings, Leda, et al. "Genomic BLAST: Custom-Defined Virtual Databases for Complete and Unfinished Genomes." *FEMS Microbiology Letters* 216, no. 2 (2002): 133–138.

7. Kohlsdorf, Daniel. "Data Mining in Large Audio Collections of Dolphin Signals." PhD diss., Georgia Institute of Technology, 2015.

8. Kershenbaum, Arik, et al. "Acoustic Sequences in Non-human Animals: A Tutorial Review and Prospectus." *Biological Reviews* 91, no. 1 (2016): 13–52.

9. Jansen, Aren, et al. "Towards Learning Semantic Audio Representations from Unlabeled Data." *Signal* 2, no. 3 (2017): 7–11.

10. Esfahanian, Mahdi, Hanqi Zhuang, and Nurgun Erdol. "On Contour-Based Classification of Dolphin Whistles by Type." *Applied Acoustics* 76 (2014): 274–279; Clink, Dena J., Abdul Hamid Ahmad, and Holger Klinck. "Brevity Is Not a Universal in Animal Communication: Evidence for Compression Depends on the Unit of Analysis in Small Ape Vocalizations." *Royal Society Open Science* 7, no. 4 (2020): 200151; Fischer, Julia, Philip Wadewitz, and Kurt Hammerschmidt. "Structural Variability and Communicative Complexity in Acoustic Communication." *Animal Behaviour* 134 (2017): 229–237.

11. McCowan, Brenda, Sean F. Hanser, and Laurance R. Doyle. "Quantitative Tools for Comparing Animal Communication Systems: Information Theory Applied to Bottlenose Dolphin Whistle Repertoires." *Animal Behaviour* 57, no. 2 (1999): 409–419.

12. McCowan, Brenda, et al. "The Appropriate Use of Zipf's Law in Animal Communication Studies." *Animal Behaviour* 69, no. 1 (2005): F1–F7.

13. Favaro, Livio, et al. "Do Penguins' Vocal Sequences Conform to Linguistic Laws?" *Biology Letters* 16, no. 2 (2020): 20190589; Watson, Stuart K., et al. "An Exploration of Menzerath's Law in Wild Mountain Gorilla Vocal Sequences." *Biology Letters* 16, no. 10 (2020): 20200380; Kershenbaum, Arik, et al. "Disentangling Canid Howls across Multiple Species and Subspecies: Structure in a Complex Communication Channel." *Behavioural Processes* 124 (2016): 149–157.

14. Hersh, Taylor A., et al. "Evidence from Sperm Whale Clans of Symbolic Marking in Non-human Cultures." *Proceedings of the National Academy of Sciences* 119, no. 37 (2022): e2201692119.

15. Sainburg, Tim, Marvin Thielk, and Timothy Q. Gentner. "Latent Space Visualization, Characterization, and Generation of Diverse Vocal Communication Signals." *bioRxiv* (2019): 870311.

16. Twiss, Sean D., and Joanna Franklin. "Individually Consistent Behavioural Patterns in Wild, Breeding Male Grey Seals (*Halichoerus grypus*)." *Aquatic Mammals* 36, no. 3 (2010): 234–238.

17. Dunn, Charlotte, et al. "Satellite-Linked Telemetry Study of a Rehabilitated and Released Atlantic Spotted Dolphin in the Bahamas Provides Insights into Broader Ranging Patterns and Conservation Needs." *Aquatic Mammals* 46, no. 6 (2020): 633–639.

18. Baird, Robin W., et al. "Movements and Habitat Use of Satellite-Tagged False Killer Whales around the Main Hawaiian Islands." *Endangered Species Research* 10 (2010): 107–121.

Chapter 7

1. Gosling, Samuel D., and Oliver P. John. "Personality Dimensions in Non-human Animals: A Cross-Species Review." *Current Directions in Psychological Science* 8, no. 3 (1999): 69–75.

2. Skrzypczak, Nathan. *Personality Traits in the Atlantic Spotted Dolphin (Stenella frontalis): Syndromes and Predictors of Neophilia.* Boca Raton: Florida Atlantic University, 2016.

3. López, Bruno Díaz. "When Personality Matters: Personality and Social Structure in Wild Bottlenose Dolphins, *Tursiops truncatus*." *Animal Behaviour* 163 (2020): 73–84.

4. Pepperberg, Irene M., and Florence A. Kozak. "Object Permanence in the African Grey Parrot (*Psittacus erithacus*)." *Animal Learning & Behavior* 14, no. 3 (1986): 322–330.

5. Savage-Rumbaugh, E. Sue, and Roger Lewin. *Kanzi: The Ape at the Brink of the Human Mind.* New York: Wiley, 1994.

6. Herman, Louis M., Douglas G. Richards, and James P. Wolz. "Comprehension of Sentences by Bottlenosed Dolphins." *Cognition* 16, no. 2 (1984): 129–219.

7. Lusseau, David. "The Emergent Properties of a Dolphin Social Network." *Proceedings of the Royal Society of London. Series B: Biological Sciences* 270, no. suppl_2 (2003): S186–S188.

8. Morton, Eugene S. "On the Occurrence and Significance of Motivation-Structural Rules in Some Bird and Mammal Sounds." *American Naturalist* 111, no. 981 (1977): 855–869.

9. McConnell, Patricia B., and Jeffrey R. Baylis. "Interspecific Communication in Cooperative Herding: Acoustic and Visual Signals from Human Shepherds and Herding Dogs." *Zeitschrift für Tierpsychologie* 67, no. 1–4 (1985): 302–328.

10. McKinley, Jean, and Thomas D. Sambrook. "Use of Human-Given Cues by Domestic Dogs (*Canis familiaris*) and Horses (*Equus caballus*)." *Animal Cognition* 3, no. 1 (2000): 13–22.

11. Filippi, Piera. "Emotional and Interactional Prosody across Animal Communication Systems: A Comparative Approach to the Emergence of Language." *Frontiers in Psychology* 7 (2016): 1393.

12. Marks, Lawrence E. *The Unity of the Senses: Interrelations among the Modalities*. New York: Academic Press, 2014.

13. King, Barbara J. *The Dynamic Dance: Nonvocal Communication in African Great Apes*. Cambridge, MA: Harvard University Press, 2009.

14. Grandin, Temple, and Catherine Johnson. *Animals in Translation: Using the Mysteries of Autism to Decode Animal Behavior*. Orlando, FL: Houghton Mifflin Harcourt, 2006.

15. Clucas, Barbara, et al. "Do American Crows Pay Attention to Human Gaze and Facial Expressions?" *Ethology* 119, no. 4 (2013): 296–302.

16. Heinrich, Bernd, and J. M. Marzluff. "Do Common Ravens Yell Because They Want to Attract Others?" *Behavioral Ecology and Sociobiology* 28 no. 1 (1991): 13–21.

17. Gardner, Howard E. *Multiple Intelligences: New Horizons in Theory and Practice*. New York: Basic Books, 2008.

18. Herzing, Denise L. "Profiling Nonhuman Intelligence: An Exercise in Developing Unbiased Tools for Describing Other 'Types' of Intelligence on Earth." *Acta Astronautica* 94, no. 2 (2014): 676–680.

19. Whitehead, Hal, and Luke Rendell. *The Cultural Lives of Whales and Dolphins*. Chicago: University of Chicago Press, 2014.

20. Pepperberg, Irene. "Animal Language Studies: What Happened?" *Psychonomic Bulletin & Review* 24, no. 1 (2017): 181–185.

21. Morton, Eugene S. "On the Occurrence and Significance of Motivation-Structural Rules in Some Bird and Mammal Sounds." *American Naturalist* 111, no. 981 (1977): 855–869.

22. McConnell, Patricia B., and Jeffrey R. Baylis. "Interspecific Communication in Cooperative Herding: Acoustic and Visual Signals from Human Shepherds and Herding Dogs." *Zeitschrift für Tierpsychologie* 67, no. 1–4 (1985): 302–328.

23. Ravindran, Sandeep. "Barbara McClintock and the Discovery of Jumping Genes." *Proceedings of the National Academy of Sciences* 109, no. 50 (2012): 20198–20199.

Chapter 8

1. Fouts, Roger S., and Joseph B. Couch. "Cultural Evolution of Learned Language in Chimpanzees." In *Communicative Behavior and Evolution*, edited by Martin E. Hahn and Edward C. Simmel, 141–161. New York: Academic Press, 1976.

2. Marino, Lori. "A Comparison of Encephalization between Odontocete Cetaceans and Anthropoid Primates." *Brain, Behavior and Evolution* 51, no. 4 (1998): 230–238.

3. Herman, Louis M., Douglas G. Richards, and James P. Wolz. "Comprehension of Sentences by Bottlenosed Dolphins." *Cognition* 16, no. 2 (1984): 129–219.

4. Savage-Rumbaugh, E. Sue, Duane M. Rumbaugh, and Sally Boysen. "Linguistically Mediated Tool Use and Exchange by Chimpanzees (*Pan troglodytes*)." In *Speaking of Apes*, edited by Thomas A. Sebeok and Jean Umiker-Sebeok, 353–383. Boston: Springer, 1980.

5. Gallup, Gordon G. "Self-Recognition in Primates: A Comparative Approach to the Bidirectional Properties of Consciousness." *American Psychologist* 32, no. 5 (1977): 329; Marino, Lori, Diana Reiss, and Gordon G. Gallup. "Mirror Self-Recognition in Bottlenose Dolphins: Implications for Comparative Investigations of Highly Dissimilar Species." In *Self-Awareness in Animals and Humans: Developmental Perspectives*, edited by Sue Taylor Parker, Robert W. Mitchell, and Maria L. Boccia, 380–391. Cambridge: Cambridge University Press, 1994; Plotnik, Joshua M., Frans B. M. De Waal, and Diana Reiss. "Self-Recognition in an Asian Elephant." *Proceedings of the National Academy of Sciences* 103, no. 45 (2006): 17053–17057.

6. Beran, Michael J., Jonathan P. Gulledge, and David A. Washburn. "Animals Count: What's Next? Contributions from the Language Research Center to Nonhuman Animal Numerical Cognition Research." In *Primate Perspectives on Behavior and Cognition*, edited by David A. Washburn, 161–173. Washington, DC: American Psychological Association, 2007.

7. Fehér, Olga. "Atypical Birdsong and Artificial Languages Provide Insights into How Communication Systems Are Shaped by Learning, Use, and Transmission." *Psychonomic Bulletin and Review* 24, no. 1 (2017): 97–105.

8. Korcsok, Beáta, et al. "Artificial Sounds following Biological Rules: A Novel Approach for Non-verbal Communication in HRI." *Scientific Reports* 10, no. 1 (2020): 1–13.

9. Heesen, Raphaela, et al. "Linguistic Laws in Chimpanzee Gestural Communication." *Proceedings of the Royal Society B* 286, no. 1896 (2019): 20182900.

10. Lyn, Heidi, Patricia Greenfield, and Sue Savage-Rumbaugh. "The Development of Representational Play in Chimpanzees and Bonobos: Evolutionary Implications, Pretense, and the Role of Interspecies Communication." *Cognitive Development* 21, no. 3 (2006): 199–213.

11. Mann, Dan C., and Marisa Hoeschele. "Segmental Units in Nonhuman Animal Vocalization as a Window into Meaning, Structure, and the Evolution of Language." *Animal Behavior and Cognition* 7, no. 2 (2020): 151–158.

12. Herzing, Denise L., and Thomas I. White. "Dolphins and the Question of Personhood." *Etica Animali* 9 (1998): 64–84.

13. Vakoch, Doug A. "In Defense of METI." *Nature Physics* 12, no. 10 (2016): 890–890.

14. Hawking, Stephen. *Into the Universe with Stephen Hawking*. Aired April 25, 2010, on Discovery Channel.

15. Delfour, Fabienne, and Ken Marten. "Inter-modal Learning Task in Bottlenosed Dolphins (*Tursiops truncatus*): A Preliminary Study Showed That

Social Factors Might Influence Learning Strategies." *Acta Ethologica* 8, no. 1 (2005): 57–64.

16. Xitco, Mark J., John D. Gory, and Stan A. Kuczaj. "Spontaneous Pointing by Bottlenose Dolphins (*Tursiops truncatus*)." *Animal Cognition* 4, no. 2 (2001): 115–123.

17. White, Thomas I. *In Defense of Dolphins: The New Moral Frontier*. Oxford: John Wiley & Sons, 2008.

18. Beguš, Gašper, et al. "Vowels and Diphthongs in Sperm Whales." OSF Preprints, December 5, 2023. https://doi.org/10.31219/osf.io/285cs.

19. Dolgin, Elie. "Elephants, Dolphins, and Chimps Need the Internet, Too: A New Initiative Promotes Internet Communication among Smart Animals-[News]." *IEEE Spectrum* 56, no. 09 (2019): 6–7.

20. McCowan, Brenda, et al. "Interactive Bioacoustic Playback as a Tool for Detecting and Exploring Nonhuman Intelligence: 'Conversing' with an Alaskan Humpback Whale." *PeerJ* 11 (2023): e16349.

21. Simões-Lopes, Paulo C., Fábio G. Daura-Jorge, and Maurício Cantor. "Clues of Cultural Transmission in Cooperative Foraging between Artisanal Fishermen and Bottlenose Dolphins, *Tursiops truncatus* (Cetacea: Delphinidae)." *Zoologia (Curitiba)* 33, no. 6 (2016).

22. Bekoff, Marc. *The Emotional Lives of Animals: A Leading Scientist Explores Animal Joy, Sorrow, and Empathy—and Why They Matter*. Novato, CA: New World Library, 2008.

23. Davis, Hank Ed, and Dianne A. Balfour. *The Inevitable Bond: Examining Scientist-Animal Interactions*. Cambridge: Cambridge University Press, 1992.

24. Kuhn, Thomas S. *The Structure of Scientific Revolutions*. Chicago: University of Chicago Press, 2012.

25. Andics, Attila, et al. "Neural Mechanisms for Lexical Processing in Dogs." *Science* 353, no. 6303 (2016): 1030–1032.

26. d'Ingeo, Serenella, et al. "Horses Associate Individual Human Voices with the Valence of past Interactions: A Behavioural and Electrophysiological Study." *Scientific reports* 9, no. 1 (2019): 1–10.

27. Card, Orson Scott. *Speaker for the Dead*. New York: Tor Books, 2009.

28. Wilson, Edward O. "Biophilia and the Conservation Ethic." In *Evolutionary Perspectives on Environmental Problems*, edited by Dustin J. Penn and Iver Mysterud, 263–272. New York: Routledge, 2017.

29. Thomas, Elizabeth Marshall. *The Tribe of Tiger: Cats and Their Culture*. New York: Simon and Schuster, 2001.

30. Bateson, Gregory. "Ecology of Mind: The Sacred." In *A Sacred Unity: Further Steps to an Ecology of Mind*. New York: Harper Collins, 1991.

31. Whitehead, Hal, et al. "Culture and Conservation of Non-humans with Reference to Whales and Dolphins: Review and New Directions." *Biological Conservation* 120, no. 3 (2004): 427–437.

32. Herzing, Denise L. *Dolphin Diaries: My 25 Years with Spotted Dolphins in the Bahamas*. New York: St. Martin's Griffin, 2011; Herzing, Denise, et al. "Imitation of Computer-Generated Sounds by Wild Atlantic Spotted Dolphins (*Stenella frontalis*)." *Animal Behavior and Cognition* 11, no. 2 (2024): 136–166. https://doi.org/10.26451/abc.11.02.02.2024.

Index